MONOGRAPHS ON
STATISTICS AND APPLIED PROBABILITY

General Editors

D.R. Cox, D.V. Hinkley, D. Rubin and B.W. Silverman

(Full details concerning this series are available from the Publishers.)

Analysis of Survival Data

D. R. COX

Nuffield College, Oxford, UK

D. OAKES

Department of Statistics,
University of Rochester, USA
formerly TUC Centenary Institute of Occupational Health
London School of Hygiene and Tropical Medicine, UK

CHAPMAN & HALL

London · Glasgow · New York · Tokyo · Melbourne · Madras

Published by Chapman & Hall, 2–6 Boundary Row, London SE1 8HN

Chapman & Hall, 2–6 Boundary Row, London SE1 8HN, UK

Blackie Academic & Professional, Wester Cleddens Road, Bishopbriggs, Glasgow G64 2NZ, UK

Chapman & Hall, 29 West 35th Street, New York NY10001, USA

Chapman & Hall Japan, Thomson Publishing Japan, Hirakawacho Nemoto Building, 6F, 1-7-11 Hirakawa-cho, Chiyoda-ku, Tokyo 102, Japan

Chapman & Hall Australia, Thomas Nelson Australia, 102 Dodds Street, South Melbourne, Victoria 3205, Australia

Chapman & Hall India, R. Seshadri, 32 Second Main Road, CIT East, Madras 600 035, India

First edition 1984
Reprinted 1985, 1988, 1990, 1992

© 1984 D.R. Cox and D. Oakes

Printed in Great Britain by the University Press, Cambridge

ISBN 0 412 24490 X

A catalogue record for this book is available from the British Library

Library of Congress Cataloging-in-Publication data
Cox, D.R. (David Roxbee)
 Analysis of survival data.
 (Monographs on statistics and applied probability)
 Bibliography: p.
 Includes indexes.
 1. Failure time data analysis. I. Oakes, D.
 II. Title. III. Series.
QA276.C665 1984 519.5 83-20882
ISBN 0-412-22490-X

Contents

Preface

The statistical analysis of the duration of life has a long history. The recent surge of interest in the topic, with its emphasis on the examination of the effect of explanatory variables, stems mainly from medical statistics but also to some extent from industrial life-testing. In fact the applications range much more widely, certainly from physics to econometrics. The essential element is the presence of a nonnegative response with appreciable dispersion and often with right censoring.

The object of the present book is to give a concise account of the analysis of survival data. We have written both for the applied statistician encountering problems of this type and also for a wider statistical audience wanting an introduction to the field.

To keep the book reasonably short we have omitted both some of the very special methods associated with the fitting of particular distributions and also the mathematically interesting topic of the application of martingale theory and weak convergence to the rigorous development of asymptotic theory. We have also firmly resisted the temptation to extend the discussion to the statistical analysis of point processes, i.e. systems in which several point events may be experienced by each individual.

We thank warmly Ms P. J. Solomon for comments on a preliminary version.

London, March 1983

D. R. Cox
D. Oakes

CHAPTER 1

The scope of survival analysis

1.1 Introduction

In survival analysis, interest centres on a group or groups of individuals for each of whom (or which) there is defined a point event, often called failure, occurring after a length of time called the failure time. Failure can occur at most once on any individual.

Examples of failure times include the lifetimes of machine components in industrial reliability, the durations of strikes or periods of unemployment in economics, the times taken by subjects to complete specified tasks in psychological experimentation, the lengths of tracks on a photographic plate in particle physics and the survival times of patients in a clinical trial.

To determine failure time precisely, there are three requirements: a time origin must be unambiguously defined, a scale for measuring the passage of time must be agreed and finally the meaning of failure must be entirely clear. We discuss these requirements in a little more detail in Section 1.2.

Sometimes we are concerned solely with the distribution of failure times in a single group. More often, we may wish to compare the failure times in two or more groups to see, for example, whether the failure times of individuals are systematically longer in the second group than in the first. Alternatively, values may be available for each individual of explanatory variables, thought to be related to survival. The lifetime of a machine component may be influenced by the stress exerted on it, or by the working temperature. White blood count is known to influence prognosis in leukaemia. In clinical practice, it is quite common for information on 100 or more variables to be routinely collected on each patient, giving the statistician the unenviable task of summarizing the joint effect of these variables on survival.

Survival analysis is properly thought of as a univariate rather than a multivariate technique because there is only a single response variable, failure time, even though there may be many explanatory variables. Some special problems involving a multivariate response are, however, discussed in Chapter 10.

1.2 The definition of failure times

We now comment briefly on the requirements for measuring failure time.

The time origin should be precisely defined for each individual. It is also desirable that, subject to any known differences on explanatory variables, all individuals should be as comparable as possible at their time origin. In a randomized clinical trial, the date of randomization satisfies both criteria, and would be the normal choice. While it might be more biologically meaningful to measure time from the first instant at which the patient's symptoms met certain criteria of severity, the difficulty of determining and the possibility of bias in such values would normally exclude their use as time origin. Such information might, however, be useful as an explanatory variable.

The time origin need not be and usually is not at the same calendar time for each individual. Most clinical trials have staggered entry, so that patients enter over a substantial time period. Each patient's failure time is usually measured from his own date of entry. Fig. 1.1 illustrates the calculation.

The evaluation of screening programmes for the detection of breast cancer provides an instructive example of the difficulties in the choice of origin. The aim of screening, of course, is to detect the disease at an earlier stage in its development than would otherwise be possible. Even in the absence of effective treatment, patients with disease detected at screening would be expected to survive longer after diagnosis than patients whose disease is detected without the aid of screening. This bias seriously complicates any comparison of the failure times of the two groups. Perhaps the only satisfactory way to evaluate the effect of screening in reducing mortality is to compare the total mortality rate in a population offered screening with that in a population where no screening programme is available.

The time origin need not always be at the point at which an individual enters the study, but if it is not, special methods are needed. For example, in epidemiological studies of the effects on mortality of occupational exposure to agents such as asbestos, the natural measure of time is age, since this is such a strong determinant of

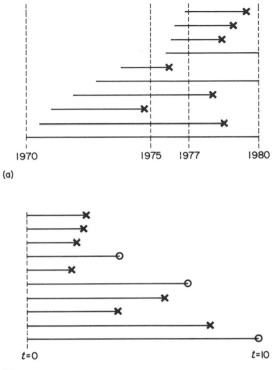

Fig. 1.1. Experience of ten individuals with staggered entry and follow-up until 1980: ×, death; ○, censoring. (a) Real time; (b) time, t, from entry into study.

mortality. However, observation on each individual commences only when he starts work in a job which involves exposure to asbestos. Likewise, in industrial reliability studies, some components may already have been in use for some period before observation begins. We refer to such data as 'left-truncated' and the appropriate methods are discussed in Chapter 11.

Often the 'scale' for measuring time is clock time (real time), although other possibilities certainly arise, such as the use of operating time of a system, mileage of a car, or some measure of cumulative load encountered. Indeed, in many industrial reliability applications, time is most appropriately measured by cumulative usage, in some sense. Or failures may consist of flaws in textile yarn, when failure 'time' would be the length measured up to the first flaw. There are interesting applications in geometrical probability, where the failure time denotes the length of a line segment contained in a convex body. About the only universal requirement for failure times is that they are nonnegative.

One reason for the choice of a timescale is direct meaningfulness for the individual concerned, justifying the use of real time in investigating survival in a medical context. Another consideration is that two individuals treated identically should, other things being equal, be in a similar state after the lapse of equal 'times'; this is the basis for the use of cumulative load encountered in an engineering context. If two or more different ways of measuring time are available, it may be possible, having selected the most appropriate timescale, to use the other 'times' as explanatory variables.

Finally, the meaning of the point event of failure must be defined precisely. In medical work, failure could mean death, death from a specific cause (e.g. lung cancer), the first recurrence of a disease after treatment, or the incidence of a new disease. In some applications there is little or no arbitrariness in the definition of failure. In others, for example in some industrial contexts, failure is defined as the first instance at which performance, measured in some quantitative way, falls below an acceptable level, defined perhaps by a specification. Then there will be some arbitrariness in the definition of failure and it will be for consideration whether to concentrate on failure time or whether to analyse the whole performance measure as a function of time.

1.3   Censoring

A special source of difficulty in the analysis of survival data is the possibility that some individuals may not be observed for the full time to failure. At the close of a life-testing experiment in industrial reliability, not all components may have failed. Some patients (many, it is to be hoped) will survive to the end of a clinical trial. A patient who has died from heart disease cannot go on to die from lung cancer.

An individual who is observed, failure-free, for 10 days and then withdrawn from study has a failure time which must exceed 10 days. Such incomplete observation of the failure time is called censoring. Note that, like failure, censoring is a point event and that the period of observation for censored individuals must be recorded.

We suppose that, in the absence of censoring, the ith individual in a sample of n has failure time T_i, a random variable. We suppose also that there is a period of observation c_i such that observation on that individual ceases at c_i if failure has not occurred by then. Then the observations consist of $X_i = \min(T_i, c_i)$, together with the indicator variable $V_i = 1$ if $T_i \leq c_i$ (uncensored), $V_i = 0$ if $T_i > c_i$ (censored). We refer to the c_i of individuals who in fact are observed to fail as unrealized censoring times, as contrasted with the realized censoring times of the censored individuals. The term potential censoring time is usual when c_i is considered without regard to whether censoring or failure occurs.

In some applications, all the c_i will be known, as for example if the only cause of censoring is the planned ending of follow-up at a predetermined time. Another example is so-called Type I censoring, in which all the c_i are equal, $c_i = c$, a constant under the control of the investigator. In Type II censoring, observation ceases after a predetermined number d of failures, so that c becomes a random variable. Type II censoring is a useful technique for economical use of effort in industrial life-testing. Other forms of so-called random censorship are possible. A crucial condition is that, conditionally on the values of any explanatory variables, the prognosis for any individual who has survived to c_i should not be affected if the individual is censored at c_i. That is, an individual who is censored at c should be representative of all those subjects with the same values of the explanatory variables who survive to c.

The simplest way to ensure this is to take the c_i to be in principle predetermined constants, and this viewpoint will be adopted throughout most of this book. Note, however, that often the c_i will not be known to the investigator in advance, and that the unrealized c_i corresponding to observed failures may never become known. The above condition is also satisfied if the potential censoring times are random variables c_i, which are independent of the T_i. Type II censoring is an example of a more general scheme in which, loosely speaking, censoring can depend on the past history, but not the future, of the whole process. We may call this evolutionary censoring.

1.4 Other methods of analysis

Besides the techniques to be discussed in this book, a number of other approaches have been used to analyse survival data. Perhaps the simplest method, much used by clinicians, is to dichotomize according to survival or nonsurvival at a critical period such as five years. Comparisons of the five-year survival rates of subjects in various groups can be made using techniques for binary data. Although this approach is often quite satisfactory, it has two major disadvantages. Concentration on a single point of the survival curve necessarily wastes some information. More seriously, calculation of survival rates as simple proportions is directly possible only when no individuals are censored during the critical period. This restriction can lead to some absurdities; see Exercise 1.1.

With survival dichotomized as above, and with quantitative explanatory variables, discriminant analysis has sometimes been used to identify variables that are related to survival, although such use of discriminant analysis is better regarded as an approach to binary logistic regression. Discriminant analysis, can, however, be a useful way of sifting through a large set of variables to determine a few variables or combinations of variables which can then be considered in more detailed analyses. By itself, discriminant analysis provides little insight into the way the explanatory variables affect survival.

Reduction to a binary response is most useful when the survival of each individual is easily classified as either very short or very long. When the potential censoring times are related to the explanatory variables, discriminant analysis will give biased results. Note also that the inclusion of the actual failure time as an explanatory variable in a discriminant analysis would be a serious error, as the failure time is part of the response, not part of the factors influencing response.

In the absence of censoring, the dependence of failure time on the explanatory variables can be explored through multiple regression. Because failure times are never negative and often have highly skewed distributions, preliminary transformations of the data such as the logarithm or reciprocal are often used. The log transformation is closely related to the accelerated life model, discussed in Chapter 5. Either transformation may give undue weight to very short failure times, which will have high negative logarithms and high positive reciprocals.

1.5   Some examples

We now describe in outline three examples that will be referred to a number of times throughout the book. Other examples will be introduced at the appropriate point in the development. Some of the examples, especially the first, have been widely used in the literature to illustrate alternative techniques.

Example 1.1   Leukaemia: comparison of two groups

Table 1.1 (Gehan, 1965, after Freireich *et al.*) shows times of remission (i.e. freedom from symptoms in a precisely defined sense) of leukaemia patients, some patients being treated with the drug 6-mercaptopurine (6-MP), the others serving as a control. Treatment allocation was randomized. Note the great dispersion and also that censoring is common in the treated group and absent in the control group. It is important to have methods of analysis that are effective in the presence of such unbalanced censoring. In fact, the trial was designed in matched pairs with one member of the pair being withdrawn from study when, or soon after, the other member comes out of remission. This is an aspect we shall ignore.

Example 1.2   Failure times and white blood count, WBC

Table 1.2 shows, for two groups of leukaemia patients, failure time (time to death) in weeks and white blood count, WBC (Feigl and Zelen, 1965). The formal difference from Example 1.1 lies partly in the presence of a continuous explanatory variable, WBC, and partly in that the division into groups is based on an (uncontrolled) measurement for each individual rather than on a randomized treatment allocation.

Example 1.3   Failure times of springs

Table 1.3 illustrates an application from industrial life-testing kindly supplied by Mr W. Armstrong. Springs are tested under cycles of repeated loading and failure time is the number of cycles to failure, it being convenient to take 10^3 cycles as the unit of 'time'. Here 60 springs were allocated, 10 to each of six different stress levels. At the lower stress levels, where failure time is long, some springs are censored, i.e. testing is abandoned before failure has occurred.

Table 1.1 *Times of remission (weeks) of leukaemia patients (Gehan, 1965, from Freireich et al.)*

Sample 0 (drug 6-MP)	6*, 6, 6, 6, 7, 9*, 10*, 10, 11*, 13, 16, 17*, 19*, 20*, 22, 23, 25*, 32*, 32*, 34*, 35*
Sample 1 (control)	1, 1, 2, 2, 3, 4, 4, 5, 5, 8, 8, 8, 8, 11, 11, 12, 12, 15, 17, 22, 23

* Censored

Table 1.3 *Cycles to failure (in units of 10^3 cycles) of springs*

Stress (N/mm²)										
950	225	171	198	189	189	135	162	135	117	162
900	216	162	153	216	225	216	306	225	243	189
850	324	321	432	252	279	414	396	379	351	333
800	627	1051	1434	2020	525	402	463	431	365	715
750	3402	9417	1802	4326	11520*	7152	2969	3012	1550	11211
700	12510*	12505*	3027	12505*	6253	8011	7795	11604*	11604*	12470*

* Censored

Table 1.2   *Failure time and white blood count (Feigl and Zelen, 1965)*

(AG positive), N = 17		(AG negative), N = 16	
White blood count, WBC	Failure time (weeks)	White blood count, WBC	Failure time (weeks)
2 300	65	4 400	56
750	156	3 000	65
4 300	100	4 000	17
2 600	134	1 500	7
6 000	16	9 000	16
10 500	108	5 300	22
10 000	121	10 000	3
17 000	4	19 000	4
5 400	39	27 000	2
7 000	143	28 000	3
9 400	56	31 000	8
32 000	26	26 000	4
35 000	22	21 000	3
100 000	1	79 000	30
100 000	1	100 000	4
52 000	5	100 000	43
100 000	65		

1.6   Computing

Some of the simpler techniques to be described in this book can be applied to modest sets of data using a programmable (or even nonprogrammable) pocket calculator. If large amounts of data are involved or if some of the more elaborate methods of analysis are contemplated, use of the computer is essential and, under the working conditions of most statisticians, the writing of special programs is impossible on other than a very small scale. Therefore, the availability of packaged programs is crucial.

All aspects of computing change so rapidly that a very detailed discussion is not appropriate in a book like this. There follow a few notes on the position at the time of writing, 1983.

The packages GLIM (Release 4), BMDP and SAS contain programs for many of the analyses described in this book. Points to watch in the choice of program include the facilities available for checking the model, e.g. through empirical survival curves, residual plots and user-defined time-dependent covariates, and the ease with which dummy variables, interactions, etc., may be incorporated in the model. A logistic regression program written by P.G. Smith (see Breslow and Day, 1980) can be used to fit the multiplicative hazards model with time-dependent covariates for small data sets.

GLM can also be used to fit some parametric models; see e.g. Aitkin and Clayton (1980) for a discussion of the Weibull distribution, in the presence of censoring. In general, methods based on the likelihood require a function maximization routine. A variety of such routines, some using derivatives of the function to be maximized, may be found in the NAG package. Ill conditioning can easily occur, particularly in attempts to discriminate between different parametric forms for the survival distribution. Routines for calculation of the complete and incomplete gamma function and its derivatives are sometimes needed; see, for example, Moore (1982), Bernardo (1976) and Schneider (1978).

Bibliographic notes, 1

A number of books on survival analysis have appeared recently. Mann *et al.* (1974), Gross and Clark (1975) and Lawless (1982) concentrate largely on fully parametric methods for particular distributions. Kalbfleisch and Prentice (1980) give a very detailed account of the multiplicative hazards model. Miller (1981) describes nonparametric and semiparametric methods. For applications in industrial reliability see Barlow and Proschan (1965, 1975), Nelson (1982) and DePriest and Launer (1983). Elandt-Johnson and Johnson (1980) describe applications in actuarial science and demography. Miké and Stanley (1982) have edited a collection of papers on medical statistics including discussion of survival data.

Armitage (1959) compared the efficiency of a number of simple methods of analysis, including the use of the proportion surviving for some specified time. Expository papers by Peto *et al.* (1976, 1977) describe the applications of some of the simpler methods for the analysis of clinical trials. Recent review papers include Prentice and Kalbfleisch (1979), Lagakos (1979) and Oakes (1981). For the mathematical theory of screening, see Prorok (1976), Shahani and Crease (1977) and Zelen and Feinleib (1969) and for an account of a large randomized trial of screening for breast cancer, see Shapiro

(1977). Three recent papers illustrating the use of survival analysis in occupational epidemiology are Liddell *et al.* (1977), Darby and Reissland (1981) and Breslow *et al.* (1983).

Further results and exercises, 1

1.1. (a) From Fig. 1.1(a) calculate the censoring times of all individuals via Fig. 1.1(b). Note that this can be done only if it is assumed that failure times can be censored solely by the conclusion of the study.

(b) The reduced sample estimator of the probability of surviving five years is the proportion, among subjects with potential censoring times exceeding five years, whose failure time is observed to exceed five years. Show that this estimator is unbiased.

(c) Show that in Fig. 1.1 the reduced sample estimators of the probabilities of surviving three years and five years are respectively $6/10$ and $4/6$. Comment.

(d) Show that if the third individual in Fig. 1.1(a) had actually entered two years earlier, but died at the same time, so that his survival would have been improved, the reduced sample estimate of the five-year survival rate for the entire group would be worsened, at $4/7$ instead of $4/6$.

1.2. Suppose that T_1, T_2 and T_3 are independent and identically distributed with a continuous distribution, and are subject to censoring times c_1, c_2 and c_3. Let $Y_i = T_i$ if $T_i \leqslant c_i$, $Y_i = \infty$ otherwise, so that Y_i may be thought of as the largest possible value of T_i consistent with the observed data. Let $X_i = \min(c_i, T_i)$. Then, on the basis of what is observed, T_1 is known to be less than or equal to T_2 if and only if $Y_1 \leqslant X_2$. Show that, whatever the values of c_1, c_2 and c_3,

(a) $\mathrm{pr}(Y_1 \leqslant X_2) = \mathrm{pr}(Y_2 \leqslant X_1),$
(b) $\mathrm{pr}(Y_1 \leqslant X_2, X_3) = \mathrm{pr}(Y_2 \leqslant X_1, X_3) = \mathrm{pr}(Y_3 \leqslant X_1, X_2),$
(c) $\mathrm{pr}(Y_1, Y_2 \leqslant X_3) = \mathrm{pr}(Y_1 < Y_3 \leqslant X_2) + \mathrm{pr}(Y_2 < Y_3 \leqslant X_1).$

[Breslow, 1970]

1.3. Suppose that data are available on a reasonably homogeneous group of patients with renal failure. All patients are initially on dialysis and the time at which they start this treatment is the time origin for each patient. All patients are observed until death. Depending on the availability of suitable donor kidneys, some patients in due course receive a kidney transplant. It is required to compare the survival under dialysis and after transplant. Criticize qualitatively the following two procedures:

(a) form two groups of patients, those never transplanted and those receiving a transplant. Compare the two distributions of time from entry to death, regardless of the time of transplant, i.e. time on dialysis of the transplanted patients is 'credited' to transplant;

(b) for the transplanted patients take a new time origin at the instant of transplant and compare the distributions of time to death for the 'dialysis only' group with that of time from transplant to death for the transplanted group.

Consider further possible procedures without the disadvantages of (a) and (b). What further difficulties are likely to arise in interpreting such data?

CHAPTER 2

Distributions of failure time

2.1 Introduction

In this chapter we consider a homogeneous population of individuals, each having a 'failure time'. That is, we deal with a single nonnegative random variable, T. In particular, an origin and scale for measuring time are assumed to be clearly defined. We examine the general specification of the distribution of T and then consider various special distributions that are useful.

We write

$$\mathscr{F}_T(t) = \mathrm{pr}(T \geqslant t) \tag{2.1}$$

for the survivor function of T, omitting the suffix T when the random variable involved is clear from the context. Mostly we deal with continuous distributions having a probability density function

$$f_T(t) = -\mathscr{F}'_T(t) = \lim_{\Delta \to 0+} \frac{\mathrm{pr}(t \leqslant T < t + \Delta)}{\Delta}, \tag{2.2}$$

so that

$$\mathscr{F}_T(t) = \int_t^\infty f_T(u)\, du.$$

Discrete and mixed discrete–continuous distributions can usually be handled formally by assigning to the probability density a component $f_j \delta(t - a_j)$ for an atom f_j at a_j, where $\delta(\,.\,)$ denotes the Dirac delta function. In the general case, the probability of survival beyond time t is the right-hand limit $\mathscr{F}(t + 0)$. Note that an unusual convention has been adopted in the definition (2.1) leading to the left continuity of the cumulative distribution function, rather than to the right continuity flowing from the standard definition. Our object is to simplify slightly some subsequent formulae involving the hazard function.

Particular forms of distribution may be useful either because they

13

are suggested by some theoretical argument or because they provide flexible empirical representations, preferably with relatively simple statistical analysis.

2.2   Hazard function

The functions $\mathscr{F}_T(.)$ and $f_T(.)$ provide two mathematically equivalent ways of specifying the distribution of a continuous nonnegative random variable, and there are of course many other equivalent functions. One with special value in the present context is the hazard function, or age-specific failure rate, defined by

$$h_T(t) = \lim_{\Delta \to 0+} \frac{\mathrm{pr}(t \leqslant T < t + \Delta \,|\, t \leqslant T)}{\Delta}. \tag{2.3}$$

By the definition of conditional probability, we have, omitting the suffix T, that

$$h(t) = f(t)/\mathscr{F}(t). \tag{2.4}$$

If there is an atom f_j of probability at time $a_j, h(t)$ contains a component $h_j \delta(t - a_j)$, where

$$h_j = f_j/\mathscr{F}(a_j), \tag{2.5}$$

and for a purely discrete distribution with atoms $\{f_j\}$ at points $\{a_j\}$, $a_1 < a_2 < \dots$,

$$h(t) = \sum h_j \delta(t - a_j),$$

where

$$\begin{aligned} h_j &= f_j/\mathscr{F}(a_j) \\ &= f_j/(f_j + f_{j+1} + \dots). \end{aligned} \tag{2.6}$$

For continuous distributions, by (2.4) and (2.2),

$$\begin{aligned} h(t) &= -\mathscr{F}'(t)/\mathscr{F}(t) \\ &= -d \log \mathscr{F}(t)/dt, \end{aligned}$$

so that, because $\mathscr{F}(0) = 1$,

$$\begin{aligned} \mathscr{F}(t) &= \exp\left(-\int_0^t h(u)\, du \right) \\ &= \exp[-H(t)], \end{aligned} \tag{2.7}$$

say, where $H(.)$ is called the integrated hazard. Further,

$$f(t) = h(t)\exp[-H(t)].\qquad(2.8)$$

If and only if $h(.)$ is constant, with value ρ say, the distribution is exponential,

$$\mathscr{F}(t) = e^{-\rho t}, \qquad f(t) = \rho e^{-\rho t}.\qquad(2.9)$$

For discrete distributions, it follows on applying (2.6) recursively, or by a direct application of the product law of probabilities, that

$$\mathscr{F}(t) = \prod_{a_j < t}(1 - h_j);\qquad(2.10)$$

to have $T \geqslant t$ it is necessary and sufficient to survive all points of support before t.

To define an integrated hazard in the discrete case the most fruitful convention is to take

$$H(t) = \sum_{a_j < t}\log(1 - h_j),\qquad(2.11)$$

so that (2.7) still holds:

$$\mathscr{F}(t) = \exp[-H(t)].$$

If the h_j are small

$$H(t) \simeq \sum_{a_j < t} h_j\qquad(2.12)$$

and the right-hand side could be taken as an alternative definition. For mixed discrete–continuous distributions, we write

$$\mathscr{F}(t) = \mathscr{P}_0^t[1 - h(u)\,du],$$

where the so-called product integral on the right-hand side is defined analogously to a Riemann integral. Divide $(0, t)$ into a large number of small intervals $[0 = x_0, x_1)$, $[x_1, x_2), \ldots, [x_{n-1}, t = x_n)$, let $\xi_j \in [x_j, x_{j+1})$ and consider the limit $n \to \infty$, with $\max(x_{j+1} - x_j) \to 0$, of

$$\prod[1 - h(\xi_j)(x_{j+1} - x_j)],$$

where $h(\xi_j)(x_{j+1} - x_j)$ is taken to be h_k if $[x_j, x_{j+1})$ contains a point of support a_k, say.

There are a number of reasons why consideration of the hazard function may be a good idea:

(i) it may be physically enlightening to consider the immediate 'risk' attaching to an individual known to be alive at age t;

(ii) comparisons of groups of individuals are sometimes most incisively made via the hazard;

(iii) hazard-based models are often convenient when there is censoring or there are several types of failure;

(iv) comparison with an exponential distribution is particularly simple in terms of the hazard;

(v) the hazard is the special form for the 'single failure' system of the complete intensity function for more elaborate point processes, i.e. systems in which several point events can occur for each individual.

2.3 Some special distributions

We now consider in outline some of the special distributions that are useful for survival data. The simpler analytical expressions of this section are summarized in Table 2.1. Of course, any distribution over nonnegative values is a possible candidate; further, any distribution, even with support including negative real values, is a possible distribution for log T.

The distributions to be discussed are all continuous. They can be classified in various ways, one being by their relation to the exponential distribution, in particular by whether they are over- or underdispersed relative to the exponential distribution.

Greek letters are used to denote adjustable parameters; ρ always has the dimensions of the reciprocal of time and can be interpreted as a rate, whereas κ and τ are dimensionless parameters. The precise interpretation of ρ, κ and τ is, however, different for the different families.

(i)   Exponential distribution

The exponential distribution of parameter ρ and mean $1/\rho$ has

$$\mathscr{F}(t) = e^{-\rho t}, \qquad f(t) = \rho e^{-\rho t}, \qquad h(t) = \rho, \qquad H(t) = \rho t.$$

The constant hazard reflects the property of the distribution reasonably called lack of memory. For any $t_0 > 0$, the conditional distribution of $T - t_0$, given $T > t_0$, is the same as the unconditional distribution of T.

The coefficient of variation, i.e. the ratio of standard deviation to

Table 2.1 *Properties of some special distributions*

		Survivor function	Density function	Hazard	No. of parameters
(i)	Exponential	$e^{-\rho t}$	$\rho e^{-\rho t}$	ρ	1
(ii)	Gamma	incomplete gamma function	$\dfrac{\rho(\rho t)^{\kappa-1} e^{-\rho t}}{\Gamma(\kappa)}$	—	2
(iii)	Weibull	$\exp[-(\rho t)^{\kappa}]$	$\kappa\rho(\rho t)^{\kappa-1}\exp[-(\rho t)^{\kappa}]$	$\kappa\rho(\rho t)^{\kappa-1}$	2
(iv)	Gompertz–Makeham	—	—	$\rho_0 + \rho_1 e^{\rho_2 t}$	3
(v)	Compound exponential	$\dfrac{(\kappa/\rho_0)^{\kappa}}{(t+\kappa/\rho_0)^{\kappa}}$	$\dfrac{\kappa(\kappa/\rho_0)^{\kappa}}{(t+\kappa/\rho_0)^{\kappa+1}}$	$\dfrac{\kappa}{t-\kappa/\rho_0}$	2
(vi)	Orthogonal polynomial	$e^{-\rho t}[1+\kappa_1\rho t+\kappa_2\,\rho t(\rho t-2)]$	$\rho e^{-\rho t}[1+\kappa_1 L_1(\rho t)+\kappa_2 L_2(\rho t)]$	—	2
(vii)	Log normal	—	—	nonmonotonic	2
(viii)	Log logistic	$[1+(t\rho)^{\kappa}]^{-1}$	$\kappa\rho^{\kappa} t^{\kappa-1}[1+(t\rho)^{\kappa}]^{-2}$	$\dfrac{\kappa t^{\kappa-1}\rho^{\kappa}}{[1+(t\rho)^{\kappa}]}$	2
(ix)	Generalized F	—	—	—	4
(x)	Inverse Gaussian	—	—	—	2
(xi)	Translation	—	—	—	1 extra for origin
(xii)	Scale family	$\mathscr{G}(\rho t)$	$\rho g(\rho t)$	$\rho h^{(\mathscr{G})}(\rho t)$	1 extra for scale
(xiii)	Proportional hazard family	$[\mathscr{L}(t)]^{\psi}$	$\psi[\mathscr{L}(t)]^{\psi-1}\,l(t)$	$\psi h^{(l)}(t)$	1 extra for proportionality

mean, is unity and this forms a reference standard for judging relative dispersion.

The exponential distribution was widely used in early work on reliability of, for example, electronic components and to a more limited extent in medical studies. That the distribution has only one adjustable parameter often means, however, that methods based on it are rather sensitive to even modest departures, for example in the tail, and the emphasis in recent work has been on methods that make less stringent assumptions about distributional form. Various idealized models lead to the exponential distribution. For example, suppose that extreme 'loads' occur in the environment in a Poisson process and that failure occurs the first time such an extreme 'load' is encountered. Again, suppose that there are many independent modes of failure, so that the observed failure time is the smallest of a large number of independent nonnegative random variables. Then, under suitable restrictions on the component random variables, the distribution of observed failure time is approximately exponential.

Note that by (2.7) if the random variable T has an arbitrary continuous distribution, then $H(T)$ has an exponential distribution with unit parameter.

We consider next a number of two-parameter families of distributions reducing to the exponential distribution by choice of one of the parameters.

(ii) Gamma distribution

The gamma family has density

$$\rho(\rho t)^{\kappa-1}e^{-\rho t}/\Gamma(\kappa), \tag{2.13}$$

where $\kappa > 0$ is an additional parameter often called the index. The mean is κ/ρ and the coefficient of variation $1/\sqrt{\kappa}$. While for many statistical purposes the gamma family is the most important family of continuous distributions taking positive values, for the present purpose the usefulness is limited by the relative clumsiness of the survivor function, an incomplete gamma integral.

The special case $\kappa = 2$ may be called the two-hit model, as it corresponds to the distribution of the time to the second point in a Poisson process of rate ρ. Other integer values of κ have an analogous interpretation.

(iii) *Weibull distribution*

The Weibull distribution with scale parameter ρ and index κ has

$$\mathcal{F}(t) = \exp[-(\rho t)^\kappa],$$
$$f(t) = \kappa\rho(\rho t)^{\kappa-1}\exp[-(\rho t)^\kappa], \qquad h(t) = \kappa\rho(\rho t)^{\kappa-1}. \qquad (2.14)$$

Because $H(t) = (\rho t)^\kappa$, it follows that T^κ has an exponential distribution of parameter ρ^κ.

The Weibull distribution arises theoretically as a limit law for the smallest of a large number of independent nonnegative random variables, thus generalizing the result already noted for the exponential distribution; see Exercise 2.7. The convenience of the Weibull distribution for empirical work stems from the simplicity of the three functions in (2.14).

(iv) *Gompertz–Makeham distribution*

A simple form of hazard function is

$$\rho_0 + \rho_1 e^{\rho_2 t} \qquad (2.15)$$

with $\rho_0 = 0$ as a special case, the Gompertz form. The associated survivor function and density follow from (2.7) and (2.8).

(v) *Compound exponential distribution*

Suppose that for each individual survival time is exponentially distributed but that the rate varies randomly between individuals. To represent this let P be a random variable with density $f_P(.)$ and suppose that the conditional density of T given $P = \rho$ is

$$f_{T|P}(t|\rho) = \rho e^{-\rho t}.$$

Then the unconditional density of T is

$$f_T(t) = \int_0^\infty \rho e^{-\rho t} f_P(\rho)\, d\rho.$$

A convenient choice for $f_P(.)$ is the gamma density of mean ρ_0 and index κ

$$f_P(\rho) = \frac{(\kappa/\rho_0)(\kappa\rho/\rho_0)^{\kappa-1} e^{-\kappa\rho/\rho_0}}{\Gamma(\kappa)}$$

leading to the Pareto distribution

$$f_T(t) = \frac{\kappa(\kappa/\rho_0)^\kappa}{(t + \kappa/\rho_0)^{\kappa+1}}. \qquad (2.16)$$

Clearly the survivor function and hazard are respectively

$$\frac{(\kappa/\rho_0)^\kappa}{(t + \kappa/\rho_0)^\kappa}, \qquad \frac{\kappa}{(t + \kappa/\rho_0)}.$$

As follows directly from its mode of construction, this distribution is overdispersed relative to the exponential distribution, to which it tends as $\kappa \to \infty$. When κ is small (2.16) has a very long tail; the rth moment exists only if $\kappa > r$.

(vi)   Expansion in orthogonal polynomials

One way of representing distributions close to a particular simple form is via an expansion in terms of orthogonal polynomials associated with the simple form. The orthogonal polynomials associated with the exponential distribution are the Laguerre polynomials and the simplest such polynomial form uses just the first- and second-degree polynomials

$$L_1(x) = x - 1, \qquad L_2(x) = x^2 - 4x + 2,$$

taking as the density

$$\rho e^{-\rho t}[1 + \kappa_1 L_1(\rho t) + \kappa_2 L_2(\rho t)]. \qquad (2.17)$$

This has mean and variance $(1 + \kappa_1)/\rho$ and $(1 + 2\kappa_1 - \kappa_1^2 + 4\kappa_2)/\rho^2$, respectively.

From one point of view (2.17) is just a mixture of an exponential distribution and gamma distributions with $\kappa = 2, 3$, all however with the same rate parameter ρ. The main role of the present expansion is in a theoretical context; when the limiting distribution for a problem is exponential, an asymptotic expansion in the form (2.17) will often result.

(vii)   Log normal distribution

As noted previously, possible distributions for T can be obtained by specifying for log T any convenient family of distributions on the real line. The simplest possibility is to take log T normally distributed with mean log ρ^{-1} and variance τ^2, leading to the log normal family for T with density

$$\frac{1}{\sqrt{(2\pi)}\tau t}\exp\left(-\frac{[\log(t\rho)]^2}{2\tau^2}\right). \tag{2.18}$$

The exponential distribution is not a special case, although a substantial amount of data is needed to discriminate empirically between an exponential distribution and a log normal distribution with $\tau \simeq 0.8$.

The hazard associated with (2.18) is nonmonotonic, although whether the maximum occurs in a range of appreciable probability depends on the value of τ. A disadvantage of the distribution for some purposes is the sensitivity of the resulting methods of statistical analysis to the small failure times.

(viii)   *Log logistic distribution*

The continuous logistic density with location v and scale parameter τ, having density

$$\frac{\tau^{-1}\exp[(x-v)/\tau]}{\{1+\exp[(x-v)/\tau]\}^2},$$

is very similar to a normal distribution. If this form is taken for log T, we obtain analogously to the log normal family, the log logistic family. It is convenient to write $v = -\log \rho$, $\kappa = 1/\tau$, so that the survivor function, density and hazard become respectively

$$\mathcal{F}(t) = \frac{1}{1+\exp[(\log t - v)/\tau]} = \frac{1}{1+(t\rho)^\kappa},$$

$$f(t) = \frac{\kappa t^{\kappa-1}\rho^\kappa}{[1+(t\rho)^\kappa]^2}, \tag{2.19}$$

$$h(t) = \frac{\kappa t^{\kappa-1}\rho^\kappa}{1+(t\rho)^\kappa}. \tag{2.20}$$

An advantage of this family over the log normal is the relatively simple explicit form achieved for $\mathcal{F}(t)$, $f(t)$ and $h(t)$. If $\kappa > 1$ the hazard has a single maximum; if $\kappa < 1$ the hazard is decreasing.

For the rth moment to exist, we need $\kappa > r$.

(ix) Comprehensive family

There are obvious technical advantages to combining distinct families of distributions into a single comprehensive family. In principle such families can always be constructed, indeed in many ways, but the result is usually too complicated to be very useful.

An occasionally useful general family in the present instance is obtained by taking T to be a multiple of the κ_1th power of a random variable $F_{(\kappa_2,\kappa_3)}$ having the standard (central) variance ratio distribution with (κ_2, κ_3) degrees of freedom. Thus

$$T = \rho^{-1} F^{\kappa_1}_{(\kappa_2,\kappa_3)}. \tag{2.21}$$

The three-parameter generalized gamma family is obtained with $\kappa_3 \to \infty$. In general, by choice of the dimensionless parameters $(\kappa_1, \kappa_2, \kappa_3)$, many of the distributions listed above can be obtained. Equation (2.21) can conveniently be written in terms of log T.

(x) Inverse Gaussian distribution

One approximate stochastic model describes failure as the first passage time of a stochastic process representing 'wear' to a fixed barrier. If the underlying process is Brownian motion with positive drift v and variance per unit time σ^2, the first passage time to a barrier at a has the inverse Gaussian distribution with density

$$\frac{a}{\sigma(2\pi t^3)^{1/2}} \exp\left(-\frac{(a-vt)^2}{2\sigma^2 t} \right). \tag{2.22}$$

This can be reparameterized in various forms, for example as

$$\left(\frac{\kappa/\rho}{2\pi t^3} \right)^{1/2} \exp\left(-\frac{\kappa\rho(t - 1/\rho)^2}{2t} \right),$$

with mean $1/\rho$ and coefficient of variation $1/\sqrt{\kappa}$, or as

$$\left(\frac{\psi}{\pi t^3} \right)^{1/2} \exp\left(-\phi t - \frac{\psi}{t} + 2\sqrt{(\phi\psi)} \right),$$

stressing the exponential family structure of the distribution.

The survivor function has the relatively complicated form

$$1 - \Phi\left[\left(\frac{\kappa}{\rho t} \right)^{1/2} (-1 + \rho t) \right] - e^{2\kappa} \Phi\left[-\left(\frac{\kappa}{\rho t} \right)^{1/2} (1 + \rho t) \right], \tag{2.23}$$

where $\Phi(.)$ is the standardized normal integral.

Although the inverse Gaussian distribution has some attractive theoretical properties and provides a reasonably flexible two-parameter family of distributions, the complexity of the survivor function makes it relatively inconvenient for handling censored data.

(xi) Translation

All the above distributions have the positive real numbers for their support. By introducing an additional parameter δ and translating the distribution, all can be converted into distributions on (δ, ∞). For instance, the translated exponential distribution has density

$$\rho e^{-\rho(t-\delta)} \qquad (t > \delta).$$

Usually we would require $\delta \geqslant 0$, although in some contexts the possibility could be contemplated of a distribution starting before the formal time origin used to define the random variable T.

(xii) Scale family

Several of the families outlined above are such that the random variable ρT has a fixed distribution, or at least a distribution involving only dimensionless shape parameters. Thus for the exponential family, ρT has the unit exponential density e^{-t}. If $\mathscr{G}(t)$, $g(t)$ and $h^{(g)}(t)$ denote respectively a survivor function, density and hazard over non-negative values, the corresponding functions

$$\mathscr{G}(t;\rho) = \mathscr{G}(\rho t),$$
$$g(t;\rho) = \rho g(\rho t), \qquad h^{(g)}(t;\rho) = \rho h^{(g)}(\rho t) \qquad (2.24)$$

define the scale family generated by $\mathscr{G}(\, . \,)$.

For the gamma family, for instance, $\mathscr{G}(\, . \,)$ depends also on κ.

Note that if $U = e^Z$ is a random variable with the density $g(\, . \,)$, then

$$\log T = -\log \rho + Z, \qquad (2.25)$$

having thus the form of a regression model.

(xiii) Lehmann family

Another useful general family is generated from the survivor function, density and hazard $\mathscr{L}(t)$, $l(t)$ and $h^{(l)}(t)$ by considering

$$\mathscr{L}(t;\psi) = [\mathscr{L}(t)]^{\psi},$$
$$l(t;\psi) = \psi[\mathscr{L}(t)]^{\psi-1} l(t), \qquad h^{(l)}(t;\psi) = \psi h^{(l)}(t). \qquad (2.26)$$

This is called the Lehmann or proportional hazards family based on $\mathscr{L}(t)$.

The families (2.24) and (2.26) are equivalent if and only if $h(t) \propto t^{\gamma}$, for some γ, so that both represent Weibull distributions; see Section 5.3 (ii).

(xiv) Qualitative analytical restrictions

A final possibility is to restrict the distribution by a qualitative analytical requirement. For example, we may require only that the hazard is nondecreasing, giving the so-called IFR (increasing failure rate) family. The analogous family, DFR, in which the hazard is nonincreasing, is rather less important.

2.4 Comparison of distributions

The families of distributions outlined above can be judged by

(i) their technical convenience for statistical inference;
(ii) the availability of explicit reasonably simple forms for survivor function, density and hazard;
(iii) the capability of representing both over- and underdispersion relative to the exponential distribution;
(iv) the qualitative shape (monotonicity, log concavity, etc.) of the hazard;
(v) the behaviour of the survivor function for small times;
(vi) the behaviour of the survivor function for large times, as judged either directly or by suitable dimensionless ratios of cumulants or moments;
(vi) any connection with a special stochastic model of failure.

Table 2.2 summarizes some of these properties in concise form. In many applications there will be insufficient data to choose between different forms by empirical analysis and then it is legitimate to make the choice on grounds of convenience; points (i) and (ii) are fairly closely related, especially when censored data are to be analysed. Behaviour for small t will be critical for some industrial applications, for instance where guarantee periods are involved, but for most medical applications the upper tail referring to relatively long survival times will be of more interest.

There are several ways of comparing different families either to

Table 2.2 Some properties useful in assessing distributional form

	$\log h(t)$	$H(t)$	$\log H(t)$	Coefficient of variation
Is it	constant? exponential	linear in t? exponential	—	1? exponential
Is it	linear in t? Gompertz $(\rho_0 = 0)$	—	linear in t? Gompertz $(\rho_0 = 0)$	<1? Gamma $(\kappa > 1)$, Weibull $(\kappa > 1)$ Log normal $(\tau < 0.83)$, Log logistic $(\tau < 0.118)$
Is it	linear in $\log t$? Weibull	—	linear in $\log t$? Weibull	—
Is it	nonmonotonic? Log normal Log logistic	—	asymptotically linear in t? Distribution with exponential tail	>1? Gamma $(\kappa < 1)$, Weibull $(\kappa < 1)$ Log normal $(\tau > 0.83)$, Log logistic $(\tau > 0.118)$ Compound exponential

highlight their differences or as a basis for an empirical analysis. On the whole, direct consideration of the density is not very effective and we concentrate here on plotting or tabulating

(a) the hazard or log hazard versus t or log t;
(b) the integrated hazard, or log survivor function, or some transform, versus t or log t;
(c) the value of the coefficient of variation, $\gamma = \sigma/\mu$;
(d) the standardized third moment $\gamma_3 = \mu_3/\sigma^3$ versus the coefficient of variation $\gamma = \sigma/\mu$, where μ, σ and μ_3 are the mean, standard deviation and third central moment of T.

Properties based on the hazard or integrated hazard lead directly to methods for analysing data that will be applicable in the presence of censoring. The integrated hazard, or log survivor function, has the advantage of indicating directly the behaviour of the upper tail of the distribution and of leading to a reasonably smooth plot when applied to empirical data. Hazard has the advantage of leading to empirical plots with points with independent errors which are, however, therefore inevitably less smooth than those of integrated hazard! It is advisable that in plots of the hazard the abscissa should be calibrated not only by t, or some function of t, but by $\mathscr{F}(t)$, so that the ranges of most concern are clear.

A value of the coefficient of variation less than one immediately excludes those distributions capable of representing only over-dispersion relative to the exponential distribution.

The graph (d) of moment ratios is particularly useful within the scale family, Section 2.3(xii), in which there is a single shape parameter. The dimensionless ratios are independent of ρ and hence the family is characterized by a curve. Fig. 2.1 shows these curves for gamma, Weibull, log normal and log logistic distributions. A high value of μ_3/σ^3, the third moment ratio, implies a relatively long tail.

The third moment has a large sampling error for long-tailed distributions and it would be possible to consider instead plots based on the standardized moments about the origin of order 3/2 or 1/2, for instance. That is with $E(T^r) = \mu'_r$, we could plot $\mu'_{3/2}/\sigma^{3/2}$ versus σ/μ or $\mu'_{1/2}/\sigma^{1/2}$ versus σ/μ; to emphasize the lower tail, we could consider $\mu'_{-1/2}\sigma^{1/2}$. The relative sensitivity of these and many other broadly similar plots is unclear.

Similar quantities for guiding the choice of models can also be calculated from log T. From censored data, comparison via the

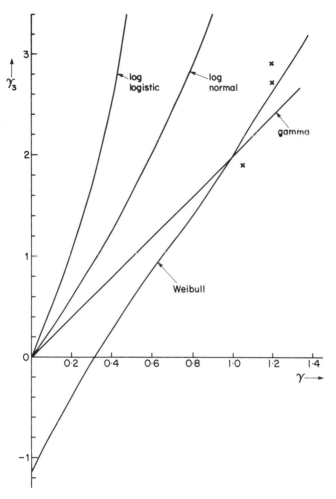

Fig. 2.1. Standardized third moment, γ_3, versus coefficient of variation, γ, for gamma, Weibull, log normal and log logistic families. Exponential distribution is at point (1, 2). ×, Boag's (1949) cancer data.

hazard or log hazard is probably the most widely useful approach, because moments cannot be calculated from censored data without strong assumptions.

Boag (1949) gave three sets of hospital cancer data and showed that the log normal distribution gives a rather better fit than the exponential. The data are quite heavily grouped and there is a little censoring. Nevertheless the coefficient of variation and standardized third moment can be estimated, attaching reasonable extrapolated values for the censored observations. The results are plotted in Fig. 2.1 and lie fairly close to the Weibull curve, showing more skewness than the gamma and less skewness than the log normal.

Finally, recall, as noted in Section 2.3(i), that one interpretation of the integrated hazard $H(.)$ is that $H(T)$ has the unit exponential distribution, i.e. $H(.)$ specifies the transformation of the timescale necessary to induce a unit exponential distribution.

Bibliographic notes, 2

The use of hazard to describe distributions of survival time has a long history in the actuarial literature. For a modern account of survival data with some emphasis on actuarial techniques, see Elandt-Johnson and Johnson (1980) and for accounts emphasizing the fitting of special distributions, see Gross and Clark (1975) and Lawless (1982).

The exponential distribution was probably studied first in connection with the kinetic theory of gases (Clausius, 1858). It plays a central role in the theory of point processes (Cox and Isham, 1980; Cox and Lewis, 1966). The Weibull distribution was introduced by Fisher and Tippett (1928) in connection with extreme value distributions; Weibull (1939a, b) studied it in an investigation of the strength of materials. Several of the other distributions are quite widely used in other statistical contexts. For the generalized F distribution, see Kalbfleisch and Prentice (1980, p. 28). Properties of the inverse Gaussian distribution are reviewed by Folks and Chhikara (1978) and Jørgensen (1982). For a summary of the properties of the main univariate continuous distributions, see Johnson and Kotz (1970). Vaupel et al. (1979) and Hougaard (1984) have examined the effect of heterogeneity between individuals via a notion of frailty.

For an account of distributions characterized by a descriptive property of the hazard, see Barlow and Proschan (1975).

Further results and exercises, 2

2.1. Suppose that a continuous random variable T is converted into a discrete random variable by grouping. Suppose that $[t - a, t)$,

$[t, t + a)$ are two adjacent groups. Examine how the values of $\mathcal{F}(t - a)$, $\mathcal{F}(t)$ and $\mathcal{F}(t + a)$ can best be used to determine $h(t)$, the underlying continuous hazard at t.

2.2. Let $T_1, \ldots, T_n$ be independent continuous nonnegative random variables with hazard functions $h_1(.), \ldots, h_n(.)$. Prove that $T = \min(T_1, \ldots, T_n)$ has hazard function $\sum h_j(t)$.

2.3. Let $T_1, \ldots, T_n$ be independent random variables with Weibull distributions with rate parameters $\rho_1, \ldots, \rho_n$ and common index κ. Prove that $T = \min(T_1, \ldots, T_n)$ also has a Weibull distribution of index κ.

2.4. In a compound exponential distribution, let the rate be represented by the random variable P. Prove that

$$E(T) = E(1/P),$$
$$\mathrm{var}(T) = 2E(1/P^2) - [E(1/P)]^2.$$

Check the results from the case where P has a gamma distribution.

2.5. Verify that special cases of the generalized F distribution of Section 2.3(ix) are achieved as follows:

$$\kappa_1 = 1, \qquad \kappa_2 = 2, \qquad \kappa_3 \to \infty, \qquad \text{exponential;}$$
$$\kappa_1 = 1, \qquad \kappa_2 \text{ arbitrary}, \qquad \kappa_3 \to \infty, \qquad \text{gamma;}$$
$$\kappa_1 \text{ arbitrary}, \qquad \kappa_2 = 2, \qquad \kappa_3 \to \infty, \qquad \text{Weibull;}$$
$$\kappa_2, \kappa_3 \to \infty, \qquad \text{log normal;}$$
$$\kappa_1 \text{ arbitrary}, \qquad \kappa_2 = \kappa_3 = 2, \qquad \text{log logistic.}$$

For the last two cases examine the moment generating function.

2.6. Prove that for the compound exponential distribution of Section 2.3(v) both density and survivor function are completely monotonic, whatever the mixing density $f_P(.)$. List some consequences.

[Widder, 1946, Chapter 4]

2.7. Suppose that $V_1, \ldots, V_m$ are independent and identically distributed continuous nonnegative random variables such that as $v \to 0$ the density and survivor function are asymptotically $av^{\kappa - 1}$ and $1 - av^\kappa/\kappa$ respectively, where $a > 0$ and $\kappa > 0$. If $W = \min(V_1, \ldots, V_m)$ and $T = (a/\kappa)^{1/\kappa} m^{1/\kappa} W$, prove that, as $m \to \infty$, T has as its limiting distribution the Weibull distribution of index κ, the exponential distribution being the limiting distribution in the special case $\kappa = 1$.

2.8. If in the representation of Exercise 2.7 the density of the V_i's is, as $v \to 0$, asymptotically $av^{\kappa-1}(1 + b_1 v + b_2 v^2 + \cdots)$, obtain expansions for the asymptotic survivor function and density of T as $m \to \infty$, showing in particular that the survivor function of T has the form

$$\begin{cases} \exp(-t^\kappa)(1 - \tfrac{1}{2}t^{2\kappa}/m + \ldots) & (\kappa < 1), \\ \exp(-t^\kappa)(1 - ct^{\kappa+1}/m^{1/\kappa} + \ldots) & (\kappa \geqslant 1), \end{cases}$$
$$c = b_1 \kappa^{1+1/\kappa}(\kappa+1)^{-1} a^{-1/\kappa} + \tfrac{1}{2}\delta_{\kappa 1},$$

with $\delta_{\kappa 1} = 1$ $(\kappa = 1)$, $\delta_{\kappa 1} = 0$ $(\kappa \neq 1)$.

Examine more general expansions for the originating density such as $at^{\kappa-1}(1 + b_1 t^\eta + \ldots)$.

2.9. If T has the Weibull distribution with parameters ρ and κ, prove that $U = \log T$ has the Gumbel distribution with survivor function $\exp(-\rho^\kappa e^{\kappa u})$ and density $\kappa \rho^\kappa \exp(\kappa u - \rho^\kappa e^{\kappa u})$. Write this in scale and location form by reparameterization. Obtain the Gumbel distribution also as the limiting distribution of $\rho \kappa T - \kappa$ as $\kappa \to \infty$.

2.10. Suppose that an individual selected at random has hazard $\rho^2 t + V$, where V is an unobserved random variable having a gamma distribution. Prove that the unconditional hazard has the form $\rho^2 t + \xi/(1 + \eta t)$, where ξ and η are parameters determined by the gamma density. Show that this can take the 'bath-tub' form with a local minimum.

[Borgefors and Hjorth, 1981]

2.11. Show that if the hazard function has the form

$$\kappa \rho (\rho t)^{\kappa-1} \exp[(\rho t)^\kappa]$$

the survivor function is

$$\exp\{-[\exp((\rho t)^\kappa) - 1]\}.$$

[Dhillon, 1979, 1981]

2.12. Prove that the square of the coefficient of variation of the log logistic distribution is $(\kappa/\pi)\tan(\pi/\kappa) - 1$, for $\kappa > 2$.

2.13. Show that the Gompertz–Makeham distribution with hazard

$$h(t) = \rho_0 + \rho_1 e^{\rho_2 t}$$

can be obtained as a compound exponential distribution provided $\rho_2 \leqslant 0$, and determine the distribution of the mixing random variable P.

2.14. Show that for the log normal distribution, the curve in Fig. 2.1 has equation

$$y = 3x + x^3,$$

where y is the standardized third moment and x is the coefficient of variation.

CHAPTER 3

Parametric statistical analysis: single sample

3.1 Introduction

In Chapter 2 we described some parametric families of survival distributions and gave some criteria for the appropriate choice of family in applications. We now suppose that a specific family has been selected, so that the distribution is known up to a vector parameter ϕ and that there is available for inference about ϕ a single sample of failure times, possibly subject to censoring. Often we may write $\phi^T = (\omega^T, \lambda^T)$, where ω is a parameter of particular interest and λ a nuisance parameter.

Here and throughout the book, we concentrate on methods based on the likelihood function. Iterative numerical solution of the likelihood equations is nearly always involved, and, as mentioned in Chapter 1, the availability of suitably flexible computer programs is crucial. After deriving the general form of the likelihood function for a censored sample, we briefly review methods of inference based on large-sample maximum likelihood theory. The exponential and Weibull distributions are considered in more detail as illustrations of the general approach. Unusually, for the exponential distribution, some 'exact' sampling theory is available.

3.2 The likelihood function

We consider first the case where the survival distribution is continuous. A subject observed to fail at t contributes a term $f(t; \phi)$ to the likelihood, the density of failure at t. The contribution from a subject whose survival time is censored at c is $\mathcal{F}(c; \phi)$, the probability of survival beyond c. The full likelihood from n independent subjects, indexed by i, is then

$$\text{lik} = \prod_{u} f(t_i; \phi) \prod_{c} \mathcal{F}(c_i; \phi), \tag{3.1}$$

where the two products are taken over uncensored and censored subjects respectively. The log likelihood is

$$l = \sum_u \log f(t_i ; \phi) + \sum_c \log \mathscr{F}(c_i ; \phi), \qquad (3.2)$$

with a similar convention for the summations.

In terms of the observed failure or censoring time $x_i = \min(t_i, c_i)$, this becomes

$$l = \sum_u \log f(x_i ; \phi) + \sum_c \log \mathscr{F}(x_i ; \phi).$$

Since $f(t) = h(t)\mathscr{F}(t)$, this may be written

$$l = \sum_u \log h(x_i ; \phi) + \sum \log \mathscr{F}(x_i ; \phi).$$

The log survivor function is minus the integrated hazard, so that

$$l = \sum_u \log h(x_i ; \phi) - \sum H(x_i ; \phi).$$

Finally, on setting $r(u) = \operatorname{card}\{i . x_i \geqslant u\}$, the number of subjects still in view at time u, we note that l may be written

$$l = \sum_u \log h(x_i ; \phi) - \int_0^\infty r(u) h(u ; \phi) \, du. \qquad (3.3)$$

Of course, the integral is only formally over an infinite range, because $r(u)$ will be zero beyond the last observed survival or censoring time. The integrand may be interpreted as the total hazard operating at time u. These expressions for l emphasize the fundamental role played by the hazard function in the development.

Suppose now that the survival distribution is discrete, with atoms $f_j(\phi)$ at preassigned points $a_j(a_1 < a_2 < \cdots)$. We shall assume that an individual censored at c could have been observed to fail at c. With this convention, the contribution to the likelihood from a subject observed to fail at a_j is $f_j(\phi)$, and from a subject censored at c is

$$\operatorname{pr}(T > c) = \mathscr{F}(c + ; \phi) = 1 - \sum_{j:a_j \leqslant c} f_j(\phi).$$

In terms of the discrete hazard function $h_j(\phi)$ given by (2.6) we have as in (2.10)

$$f_j(\phi) = h_j(\phi) \prod_{k < j} [1 - h_k(\phi)],$$

$$\mathscr{F}(c + ; \phi) = \prod_{j:a_j \leqslant c} [1 - h_j(\phi)].$$

Each term is a product over the atoms $\{a_j\}$ of the survival distribution.

To derive the full likelihood from a sample of n observations, we first collect all the terms corresponding to the atom a_j. If there are d_j failures among the $r_j = r(a_j)$ individuals in view at a_j, the contribution to the total likelihood is

$$[h_j(\phi)]^{d_j}[1 - h_j(\phi)]^{r_j - d_j}.$$

The total log likelihood is then

$$\sum_j \{d_j \log h_j(\phi) + (r_j - d_j) \log [1 - h_j(\phi)]\}. \tag{3.4}$$

Note that this is the same as would be obtained from a series of independent binomial terms, with r_j trials and probability of 'success' $h_j(\phi)$.

In practice, truly discrete survival distributions are rarely encountered. Ties in reported values are usually due to the grouping of data from an underlying continuous distribution. For most purposes, and especially for the single-sample problem, the consequent small inaccuracies in the data can generally be ignored. An exception to this rule is the fitting of a log normal distribution to data with many values close to zero. As noted in Chapter 2, the fitted parameters are sensitive to the very short survival times.

The exact likelihood for grouped data can be derived: it involves integrals of the density function over the grouping intervals.

3.3   Likelihood theory: general considerations

Various approaches are possible to the extraction of information about ϕ from the log likelihood function (3.2) or (3.4). If a prior distribution is available for the unknown parameter the usual calculations of Bayesian theory lead to the posterior distribution of the parameter of interest. Note that in the case $\phi^T = (\omega^T, \lambda^T)$, a joint prior distribution is needed over the parameter of interest ω and the nuisance parameter λ. If a sampling theory approach is used, it may be possible to develop 'exact' confidence intervals and tests, perhaps eliminating the nuisance parameter by a conditioning argument.

More commonly, the asymptotic considerations of maximum likelihood theory are used. Three broad types of asymptotic procedure, based on likelihood, are available for testing the null hypothesis $\omega = \omega_0$ and hence for deriving a confidence set for ω as the collection of parameter values not 'rejected' at the level in question. These types are as follows:

(a) First, there is the direct use of the likelihood ratio statistic

$$W(\omega_0) = W = 2[l(\hat{\omega}, \hat{\lambda}) - l(\omega_0, \hat{\lambda}_{\omega_0})], \qquad (3.5)$$

where $(\hat{\omega}, \hat{\lambda})$ is the joint maximum likelihood estimate of (ω, λ) and $\hat{\lambda}_{\omega_0}$ is the maximum likelihood estimate of λ when $\omega = \omega_0$. The function $l(\omega, \hat{\lambda}_\omega)$ is sometimes called the profile log likelihood for ω. Under the null hypothesis $\omega = \omega_0$, $W(\omega_0)$ has, approximately, a chi-squared distribution with $p_\omega = \dim(\omega)$ degrees of freedom. The corresponding $1 - \alpha$ confidence region is

$$\{\omega : W(\omega) \leqslant c^*_{p_\omega, \alpha}\}, \qquad (3.6)$$

where $c^*_{p, \alpha}$ is the upper α point of the chi-squared distribution with p degrees of freedom. If the asymptotic distribution were exact, we would have $E[W(\omega_0); \omega_0] = p_\omega$. Often it is possible to find an expansion

$$E[W(\omega_0); \omega_0] = p_\omega \left[1 + \frac{c}{n} + o\left(\frac{1}{n}\right) \right].$$

Then $(1 + c/n)$, with if necessary c estimated consistently, is called a Bartlett correction factor and improved properties are obtained by replacing W by

$$W' = W/(1 + c/n)$$

in (3.5) and (3.6). It is rarely feasible, however, to carry out such calculations in the presence of censoring.

(b) Secondly, we may make direct use of the maximum likelihood estimate $\hat{\omega}$. The observed information matrix is the matrix of minus the second derivatives of l with respect to (ω, λ), evaluated at $(\hat{\omega}, \hat{\lambda})$. Write $v_{\omega\omega}(\hat{\omega}, \hat{\lambda})$ for the leading submatrix of the inverse of the observed information matrix; it can be regarded as the estimated covariance matrix of $\hat{\omega}$. Then we may use instead of (3.5) the Wald statistic

$$W_e(\omega_0) = (\hat{\omega} - \omega_0)^T v_{\omega\omega}^{-1}(\hat{\omega}, \hat{\lambda})(\hat{\omega} - \omega_0), \qquad (3.7)$$

again with an approximate chi-squared distribution with p_ω degrees of freedom under the null hypothesis. Equation (3.7) leads directly to an elliptical confidence region for ω, centred on $\hat{\omega}$. There are alternative ways of estimating the covariance matrix, for example via expected rather than observed second derivatives of the log likelihood. If ω is a scalar parameter, there results the symmetric $1 - 2\alpha$ confidence interval

$$\hat{\omega} - k_\alpha^* v_{\omega\omega}^{1/2}(\hat{\omega},\hat{\lambda}), \qquad \hat{\omega} + k_\alpha^* v_{\omega\omega}^{1/2}(\hat{\omega},\hat{\lambda}),$$

where $\Phi(-k_\alpha^*) = \alpha$.

(c) A third possibility is to use the gradient of the log likelihood at ω_0, replacing λ by $\hat{\lambda}_{\omega_0}$, i.e. to calculate

$$U_{\omega_0} = \left[\frac{\partial}{\partial\omega}l(\omega,\lambda)\right]_{\omega=\omega_0,\,\lambda=\hat{\lambda}_{\omega_0}}. \tag{3.8}$$

This $p_\omega \times 1$ vector is, when $\omega = \omega_0$, approximately normally distributed with zero mean and covariance matrix $v_{\omega\omega}^{-1}(\omega_0,\hat{\lambda}_{\omega_0})$. The test statistic based on U_{ω_0} is

$$W_U(\omega_0) = U_{\omega_0}^T v_{\omega\omega}(\omega_0,\hat{\lambda}_{\omega_0})U_{\omega_0}. \tag{3.9}$$

Again there are alternative ways of evaluating the covariance matrix, and the distribution under the null hypothesis is approximately chi-squared with p_ω degrees of freedom.

We note in passing that the estimated covariance matrix, evaluated at $\omega = \omega_0$, $\lambda = \hat{\lambda}_0$, of

$$\left[\frac{\partial}{\partial\omega}l(\omega,\lambda)\right]$$

is $I_{\omega\omega}(\omega_0,\hat{\lambda}_{\omega_0})$, the leading submatrix of the observed information matrix. In general, $I_{\omega\omega}(\omega_0,\hat{\lambda}_{\omega_0}) \neq v_{\omega\omega}^{-1}(\omega_0,\hat{\lambda}_{\omega_0})$; the difference represents the gain in information about ω provided by knowledge of λ.

The three procedures (a)–(c) will very often give virtually identical answers. Procedure (b), the direct use of the maximum likelihood estimate, has advantages in simple presentation of conclusions, but the disadvantage is that it is not invariant under reparameterization and that it may yield absurd answers if the likelihood is of unusual shape, e.g. is multimodal or zero for certain values of ϕ. Procedure (c) has some computational advantage in that only maximization at $\phi = \phi_0$ is required, so that the adequacy of a basic model specified by λ can be tested by augmentation in various directions without remaximization. In cases of doubt, procedure (a), the direct use of maximized log likelihoods is recommended. It is invariant under reparameterization and the shape of the resulting confidence region is settled by the data. These qualitative arguments are reinforced by recent work on higher-order asymptotic theory, in particular involving conditioning on approximately ancillary statistics (Efron and Hinkley, 1978; Cox, 1980; Barndorff-Nielsen, 1980, 1983; Barndorff-Nielsen and Cox, 1984).

The asymptotic theory of maximum likelihood estimation on which the normal and chi-squared approximations are based does require the satisfaction of some 'regularity conditions' concerning the smoothness of the likelihood function. In particular, the theory does not hold for threshold parameters. With observations that are not independent or identically distributed, it is necessary, roughly speaking, that the proportion of the total information in the sample contributed by any single observation should converge to zero as the sample size increases. Of course, even if the problem is such that the conditions for asymptotic normality hold, the theory may give a poor approximation to the small-sample results.

If there is serious doubt about the applicability of asymptotic distribution theory, a Bartlett correction factor can be calculated or computer simulation used to examine the distribution on the null hypothesis of any appropriate statistic.

3.4   Exponentially distributed failure times

The exponential distribution, with rate parameter ρ, $\mathscr{F}(t) = e^{-\rho t}$, has constant hazard function $h(t, \phi) = h(t, \rho) = \rho$. The log likelihood for the single unknown parameter ρ is thus

$$l = \sum_u \log \rho - \rho \sum x_i = d \log \rho - \rho \sum x_i, \qquad (3.10)$$

say, and the total number d of failures and the total $\sum x_i$ of the censored and uncensored failure times form a minimal sufficient statistic for ρ. Note that unless d or $\sum x_i$ or some function of them is fixed by design, we have a two-dimensional statistic for a one-dimensional parameter, showing an example of a so-called curved exponential family. Often $\sum x_i$ is called the total time at risk.

The derivatives of l are

$$U_\rho = \partial l / \partial \rho = d / \rho - \sum x_i, \qquad (3.11)$$

$$I = -\partial^2 l / \partial \rho^2 = d / \rho^2. \qquad (3.12)$$

The maximum likelihood estimator $\hat{\rho}$ of ρ is the solution of $U_\rho = 0$, namely

$$\hat{\rho} = d / \sum x_i, \qquad (3.13)$$

the total number of failures divided by the total time at risk. Censored failure times contribute to the denominator but not to the numerator of this ratio.

When there is no censoring, the log likelihood becomes

$$l = n \log \rho - \rho \sum x_i,$$

and the curved exponential family collapses to a full one-dimensional family with the single minimal sufficient statistic $\sum x_i$ for ρ. Here, exact inference for ρ is possible, because $\sum x_i$, the sum of n independent exponentially distributed random variables with the same parameter ρ, has a gamma distribution with index n and scale parameter ρ. Thus $2n\rho/\hat{\rho}$ has a chi-squared distribution with $2n$ degrees of freedom. Interval estimates and hypothesis tests for ρ follow immediately. In particular, a $1 - \alpha$ confidence interval for ρ is

$$\frac{\hat{\rho} c^*_{2n, 1 - \frac{1}{2}\alpha}}{2n} < \rho < \frac{\hat{\rho} c^*_{2n, \frac{1}{2}\alpha}}{2n},$$

where $c^*_{p,\alpha}$ is the upper α point of the chi-squared distribution with p degrees of freedom.

Example 3.1

Consider the leukaemia data of Freireich et al. given in Table 1.1. For the control group, with no censoring, $n = 21$ and $\sum x_i = 182$. If an exponential distribution is assumed

$$\hat{\rho} = 21/182 = 0.115,$$

and an exact 95% confidence interval for ρ is (0.071, 0.170), since the 0.025 and 0.975 points of the chi-squared distribution with 42 degrees of freedom are respectively 26.0 and 61.8.

The exact theory holds also with Type II censoring, that is when observation ceases after a predetermined number of failures, d. This is easily seen by noting that if $t_{(i)}$ denotes the ith ordered failure time ($i = 0, 1, \ldots, d$: $t_{(0)} = 0$), then $t_{(i)} - t_{(i-1)}$ has an exponential distribution with parameter $(n - i + 1)\rho$ and that

$$\sum_{i=1}^{n} x_i = \sum_{i=1}^{d} (n - i + 1)(t_{(i)} - t_{(i-1)}),$$

with this censoring mechanism.

With other censoring patterns, the exact sampling distribution of $\hat{\rho}$ is difficult to derive. It has been tabulated for the special case of Type I censoring, when observation on all individuals ceases at a predetermined time c. However, good approximate procedures for general censoring patterns can be obtained by treating $2d\rho/\hat{\rho}$ as a chi-squared variable on $2d$ degrees of freedom, ignoring the fact that d is now a random variable. The resulting confidence intervals are very similar to those obtained from the likelihood ratio.

Example 3.1 (continued)

For the treated group (6-MP), $\sum x_i = 359$, $d = 9$. The maximum likelihood estimator is

$$\hat{\rho} = d/\sum x_i = 9/359 = 0.025.$$

The log likelihood function is plotted in Fig. 3.1 and shows a noticeable lack of symmetry. The 95% confidence interval for ρ from the likelihood ratio is (0.0120, 0.0452). The interval obtained from the upper and lower 0.025 points of the chi-squared distribution with 18

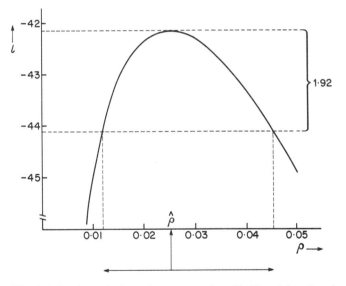

Fig. 3.1. Leukaemia data, 6-MP group. Log likelihood function, l, for exponential parameter, ρ. 95% confidence interval derived from chi-squared distribution. Maximum likelihood estimate, $\hat{\rho}$.

degrees of freedom is $(0.0115, 0.0439)$. The standard error of the maximum likelihood estimate is

$$\left(\left[-\frac{\partial^2 l}{\partial \rho^2}\right]_{\hat{\rho}}\right)^{-1/2} = \left(\frac{\hat{\rho}^2}{d}\right)^{1/2} = 0.00836.$$

This gives a symmetric 95% confidence interval, based on a normal approximation to the distribution of $\hat{\rho}$, of $(0.0087, 0.0414)$. In view of the shape of the likelihood function, this symmetric interval would be an inappropriate choice here.

3.5 Proportional hazards family

The likelihood equations take a simple form for the Lehmann family $h(t;\rho) = \rho h_0(t)$, where the hazard function $h_0(t)$ is assumed known. In fact, if

$$H_0(t) = \int_0^t h_0(u)\, du$$

denotes the integrated hazard corresponding to $h_0(t)$, the random variables $T_i' = H_0(T_i)$ will have exponential distributions with parameter ρ, so that the methods of the previous section apply.

Alternatively, we may proceed directly from the log likelihood. For a general censored sample, equation (3.3) gives

$$l = d \log \rho + \sum_u \log h_0(x_i) + \rho e, \tag{3.14}$$

where

$$e = \int_0^\infty r(u) h_0(u)\, du = \sum_{i=1}^n H_0(x_i).$$

It is easily seen that, whatever the censoring mechanism, the random variable ρe has the same expectation as the number d of observed failures; see Exercise 3.2. The derivative of l is

$$dl/d\rho = d/\rho - e, \tag{3.15}$$

leading to the simple form

$$\hat{\rho} = d/e \tag{3.16}$$

for the maximum likelihood estimator. In epidemiological applications, the $h_0(t)$ may represent known age-specific mortality rates for a given 'standard population', and d the number of deaths observed in a 'study population' of interest. The ratio d/e, possibly expressed as a percentage, is called the standardized mortality ratio (SMR). It is usually necessary to allow for dependence of the hazard function on calendar time as well as age, in the calculations of e.

The $1 - \alpha$ 'limits on the expectation' of a Poisson variable with observed value k, are

$$\theta_L(k) = \bar{\theta}(k, \tfrac{1}{2}\alpha), \qquad \theta_U(k) = \bar{\theta}(k + 1, 1 - \tfrac{1}{2}\alpha),$$

where $\bar{\theta}(k, \alpha)$ is the root of

$$\sum_{i=k}^{\infty} \frac{\theta^i}{i!} \exp(-\theta) = \alpha. \tag{3.17}$$

A $1 - \alpha$ confidence interval for ρ can be calculated from these limits as $(\theta_L(d)/e, \theta_U(d)/e)$, i.e. treating d as if it had a Poisson distribution with mean ρe, with e nonrandom. It can be shown that the lower limit of this interval is the same as that obtained by taking $2\rho e$ to have a chi-squared distribution with $2d$ degrees of freedom. For the upper limit, the degrees of freedom, however, must be taken as $2d + 2$.

3.6 Likelihood estimation for the Weibull distribution

We now consider maximum likelihood estimation for the parameters (κ, ρ), both assumed unknown, of the Weibull distribution, with hazard $h(t) = \kappa\rho(\rho t)^{\kappa - 1}$. From (3.2) the log likelihood from a censored sample is

$$l = d \log \kappa + \kappa d \log \rho + (\kappa - 1)\sum_u \log x_i - \rho^\kappa \sum x_i^\kappa.$$

Even in the absence of censoring, there is no fixed dimensional sufficient statistic for (ρ, κ); the Weibull is not an exponential family. The first derivatives are

$$U_\rho = \frac{\partial l}{\partial \rho} = \frac{\kappa d}{\rho} - \kappa\rho^{\kappa - 1} \sum x_i^\kappa, \tag{3.18}$$

$$U_\kappa = \frac{\partial l}{\partial \kappa} = \frac{d}{\kappa} + d \log \rho + \sum_u \log x_i - \rho^\kappa \sum x_i^\kappa \log(\rho x_i). \tag{3.19}$$

If κ is specified, the maximum likelihood estimator $\hat{\rho}_\kappa$ of ρ can be found explicitly by solving $U_\rho = 0$ as

$$\hat{\rho} = (d/\textstyle\sum x_i^\kappa)^{1/\kappa}, \tag{3.20}$$

a result which could be derived immediately from the fact that T^κ has an exponential distribution with parameter ρ^κ. Substitution into the equation $U_\kappa = 0$ yields the simpler form

$$0 = \frac{d}{\kappa} + \sum_u \log x_i - d\frac{\sum x_i^\kappa \log x_i}{\sum x_i^\kappa} \tag{3.21}$$

for the maximum likelihood estimator $\hat{\kappa}$. Equation (3.21) does not contain ρ and can be solved by a one-dimensional iterative scheme in κ.

The second derivatives of l are

$$-I_{\rho\rho} = \frac{\partial^2 l}{\partial \rho^2} = -\frac{\kappa d}{\rho^2} - \kappa(\kappa - 1)\rho^{\kappa-2}\sum x_i^\kappa, \tag{3.22}$$

$$-I_{\rho\kappa} = \frac{\partial^2 l}{\partial \rho \partial \kappa} = \frac{d}{\rho} - \rho^{\kappa-1}(1 + \kappa \log \rho)\sum x_i^\kappa - \kappa\rho^{\kappa-1}\sum x_i^\kappa \log x_i,$$
$$\tag{3.23}$$

$$-I_{\kappa\kappa} = \frac{\partial^2 l}{\partial \kappa^2} = -\frac{d}{\kappa^2} - \rho^\kappa\sum x_i^\kappa[\log(\rho x_i)]^2. \tag{3.24}$$

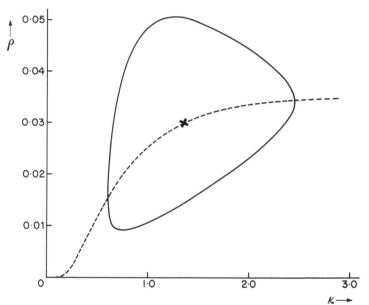

Fig. 3.2. Leukaemia data, 6-MP group. Fitting of Weibull distribution: ———, boundary of 95% confidence region based on likelihood ratio statistic; –––––, maximum likelihood estimate $\hat{\rho}_\kappa$ of ρ for given κ; ×, maximum likelihood estimate $(\hat{\kappa}, \hat{\rho})$.

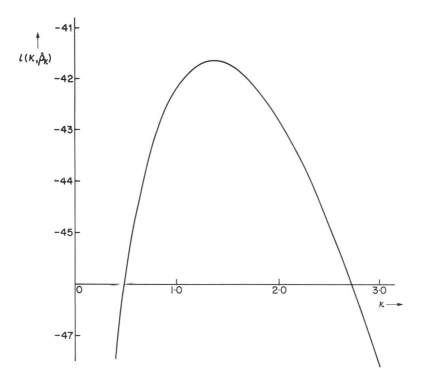

Fig. 3.3. Leukaemia data, 6-MP group. Log likelihood profile for Weibull parameter, κ, i.e. $l(\kappa, \hat{\rho}_\kappa)$ versus κ.

Example 3.2

For the leukaemia data (treated group) the joint maximum likelihood estimator of (κ, ρ) is $\hat{\kappa} = 1.35$, $\hat{\rho} = 0.030$. Fig. 3.2 shows the joint 95% confidence region for (κ, ρ) obtained from the likelihood ratio statistic W. For given κ the maximum likelihood estimator $\hat{\rho}_\kappa$ of ρ is also shown as a function of κ. Fig. 3.3 shows the likelihood profile $l(\kappa, \hat{\rho}_\kappa)$. The 95% confidence interval for κ based on W is (0.72, 2.20).

3.7 A test for exponentiality

We now derive the score test of the hypothesis $\kappa = 1$ corresponding to exponentiality. This will be a useful test against alternative hypotheses which specify monotone hazard functions.

The maximum likelihood estimator $\hat{\rho}_\kappa$ of ρ when $\kappa = \kappa_0 = 1$ is just

$$\hat{\rho}_{\kappa_0} = d / \sum x_i.$$

The score function is, from (3.19),

$$U_{\kappa_0} = \left[\frac{\partial}{\partial \kappa} l(\kappa, \rho) \right]_{\kappa_0, \rho = \hat{\rho}_{\kappa_0}}$$

$$= d + \sum_u \log x_i - d \frac{\sum x_i \log x_i}{\sum x_i}.$$

The observed information matrix at $(\kappa_0, \hat{\rho}_{\kappa_0})$ has elements

$$I_{\kappa\kappa} = d + \sum (\hat{\rho}_{\kappa_0} x_i) [\log(\hat{\rho}_{\kappa_0} x_i)]^2,$$
$$I_{\kappa\rho} = \sum x_i \log(\hat{\rho}_{\kappa_0} x_i),$$
$$I_{\rho\rho} = d / \hat{\rho}_{\kappa_0}^2.$$

The inverse matrix v has leading element

$$v_{\kappa\kappa} = (I_{\kappa\kappa} - I_{\kappa\rho}^2 / I_{\rho\rho})^{-1},$$

and the approximate chi-squared statistic can be constructed as at (3.9). When, as here, $p_\omega = 1$, the signed statistic

$$U_{\kappa_0}(v_{\kappa\kappa})^{1/2},$$

approximately a standard normal deviate on the null hypothesis, is to be preferred, as it indicates the direction of the departure from the null hypothesis.

Example 3.3

For the leukaemia data,

$$d = 9, \qquad \sum x_i = 359, \qquad \sum_u \log x_i = 21.19,$$
$$\sum x_i \log x_i = 1077.3, \qquad \sum x_i (\log x_i)^2 = 3334.8.$$

Thus, $\hat{\rho}_{\kappa_0} = 0.02507$ and $U_{\kappa_0} = 3.18$. The elements of the observed information matrix at $\kappa = \kappa_0$ are

$$I_{\kappa\kappa} = 15.79, \qquad I_{\kappa\rho} = -246.0, \qquad I_{\rho\rho} = 14320,$$

and $v_{\kappa\kappa} = 0.0865$. The standard normal deviate for the test of $\kappa = \kappa_0$ is $3.18 \times (0.0865)^{1/2} = 0.935$, indicating consistency with the null hypothesis of exponentiality. This agrees qualitatively with the conclusions from the likelihood ratio statistic. In view of the small number of failures, it is not surprising that the null hypothesis cannot be rejected. The evidence, weak as it is, in favour of a monotone increasing rather than a monotone decreasing hazard, is probably due to the apparent threshold at six weeks.

Bibliographic notes, 3

Estimation from censored samples when the number of failures is predetermined and the distribution exponential was considered by Sukhatme (1937) and Epstein and Sobel (1953). Bartholomew (1957) derived asymptotic methods for the exponential distribution with fixed censoring times and also (Bartholomew, 1963) gave the exact distribution of the maximum likelihood estimator for Type I censoring. Cox (1953) suggested a chi-squared approximation in the context of a single Poisson process observed for a fixed time.

The computation and interpretation of standardized mortality ratios is discussed in most texts on medical statistics; see, for example, Bradford Hill (1977). Breslow (1977) gave the likelihood derivation. Limits on the expectation of a Poisson variable are tabulated by Fisher and Yates (1963) and Pearson and Hartley (1966). The explicit connection with chi-squared was noted by Fisher (1935) but was known much earlier, in a different context, to A. K. Erlang.

Likelihood estimation in the Weibull distribution was discussed in detail by Pike (1966) and Peto and Lee (1973). The reliability literature (Mann *et al.*, 1974) contains many alternative procedures. These are often based on order statistics and are thus applicable in the presence of a threshold parameter, as well as of censoring.

For a general introduction to large-sample likelihood theory, see Rao (1973, Chapter 6) and Cox and Hinkley (1974, Chapter 9).

Further results and exercises, 3

3.1. Suppose that failure times are exponentially distributed with parameter ρ, and that ρ has a prior distribution of the gamma form $f_P(\rho)$ of Section 2.3. Show that the posterior distribution of ρ given the number of observed failures d and total time at risk $\sum x_i$ is also gamma, with parameters $\kappa_1 = \kappa + d$, $\rho_1 = (\kappa + d)\rho_0/(\kappa + \rho_0 \sum x_i)$. Consider the extension to the proportional hazards family, with $h_0(t)$ assumed known.

3.2. Show that, if T has a continuous cumulative distribution function F with integrated hazard $H(\,.\,)$, and if $X = \min(T, c)$, then $E[H(X)] = F(c)$.

3.3. Show that, if there are at least two distinct uncensored failure times, and each $x_i > 0$, equation (3.21) always has a unique root in $\kappa > 0$.

3.4. Show that in the absence of censoring the expected information matrix for the Weibull distribution has elements

$$-E\left(\frac{\partial^2 l}{\partial \rho^2}\right) = \frac{n\kappa^2}{\rho^2}, \qquad -E\left(\frac{\partial^2 l}{\partial \rho \partial \kappa}\right) = \frac{n\psi(2)}{\rho},$$

$$-E\left(\frac{\partial^2 l}{\partial \kappa^2}\right) = \frac{n}{\kappa^2}\{1 + \psi'(2) + [\psi(2)]^2\},$$

where

$$\psi(\alpha) = \frac{\partial}{\partial \alpha} \log \Gamma(\alpha)$$

is the digamma function.

3.5. Show that in the absence of censoring, the gamma distribution (2.13) has minimal sufficient statistic $(\sum t_i, \sum \log t_i)$ for ρ and κ, and that the maximum likelihood estimator of κ with unknown ρ is the solution of

$$\psi(\kappa) - \log \kappa - \log R = 0,$$

where R is the ratio of the geometric to the arithmetic mean of the sample.

3.6. By considering (i) the number of events and (ii) the time to the kth event in a Poisson process of rate θ observed over the unit interval, show that

$$\bar{\theta}(k, \alpha) = \tfrac{1}{2} c^*_{2k, \alpha}.$$

3.7. For n uncensored individuals with exponentially distributed failure times and log likelihood function $n \log \rho - \rho \sum x_i$, prove that the likelihood ratio statistic W of (3.5) for testing the null hypothesis $\rho = \rho_0$ is

$$W = 2(n \log n - n \log \textstyle\sum x_i - n - n \log \rho_0 + \rho_0 \sum x_i).$$

Prove that the expected value under the null hypothesis is

$$E(W) = 1 + (6n)^{-1} + O(n^{-2}),$$

so that the statistic with a Bartlett correction factor is

$$W' = W/[1 + (6n)^{-1}].$$

Examine numerically the relation between confidence limits from W, from W', and from the 'exact' solution.

CHAPTER 4

Single-sample nonparametric methods

4.1 Introduction

The methods of the previous chapter all require specification of the functional form of the distribution that failure time would have in the absence of censoring. We now discuss nonparametric techniques which require no such assumptions. As well as providing flexible alternatives to the parametric techniques, they are useful in connection with graphical assessments of goodness of fit for complex models. The term 'life table' is often used for a nonparametric estimate of a survivor function from censored data.

4.2 Product-limit estimator

We begin by assuming that the possibly improper distribution is discrete, with atoms f_j at finitely many specified points $a_1 < a_2 < \cdots < a_g$. In practice, these points are often taken to be equally spaced, $a_j = j$ in suitable time units, but this is not necessary. As described in Section 2.2, the survivor function $\mathscr{F}(t)$ may be expressed in terms of the discrete hazard function h_j as

$$\mathscr{F}(t) = \prod_{a_j < t} (1 - h_j) = \prod{}^{(t)}(1 - h_j),$$

where $\prod{}^{(t)}$, and subsequently $\sum{}^{(t)}$, denote product and sum over j, $a_j < t$. Thus, in terms of the h_j, the f_j may be written in the form

$$f_1 = h_1, \qquad f_2 = (1 - h_1)h_2, \qquad \ldots,$$
$$f_j = (1 - h_1)(1 - h_2)\ldots(1 - h_{j-1})h_j, \qquad \ldots,$$
$$f_g = (1 - h_1)(1 - h_2)\ldots(1 - h_{g-1})h_g. \tag{4.1}$$

The constraints $f_j \geqslant 0$, $\sum f_j \leqslant 1$ become, simply, $0 \leqslant h_j \leqslant 1$.

A nonparametric estimator of the survivor function is

$$\hat{\mathscr{F}}(t) = \prod^{(t)}(1 - \hat{h}_j), \tag{4.2}$$

where the $\hat{h}_j$ are the maximum likelihood estimators of the h_j. From equation (3.4), the log likelihood in terms of the h_j is

$$\sum_j [d_j \log h_j + (r_j - d_j)\log(1 - h_j)], \tag{4.3}$$

where r_j is the number of individuals in view at a_j, and d_j is the number who fail at a_j. It is conventional to include in r_j any individuals who are censored at a_j. Other conventions are possible and would lead to slightly different results in the sequel. Any difficulty that this causes in practice can usually be resolved by obtaining the data recorded on a finer scale.

The log likelihood (4.3) is exactly that for g independent binomials, with respectively r_j trials, d_j failures, and probability of failure h_j. It is particularly easy to maximize here, as the parameter vector is $\{h_j\}$ itself. Thus, $\hat{h}_j$ is the solution of

$$\frac{\partial l}{\partial h_j} = \frac{d_j}{h_j} - \frac{r_j - d_j}{1 - h_j} = 0,$$

i.e. $\hat{h}_j = d_j/r_j$. The corresponding estimator $\hat{\mathscr{F}}$ of the survivor function is

$$\hat{\mathscr{F}}(t) = \prod^{(t)}\left(1 - \frac{d_j}{r_j}\right), \tag{4.4}$$

obtained by substituting in (4.2).

Table 4.1 *Calculation of the product-limit estimator for the leukaemia data of Table 1.1, 6-MP group*

a_j	r_j	d_j	$1 - \dfrac{d_j}{r_j}$	$\prod\limits_{l \leqslant j}\left(1 - \dfrac{d_l}{r_l}\right) = \hat{\mathscr{F}}(a_j +)$
6	21	3	0.8571	0.8571
7	17	1	0.9412	0.8067
10	15	1	0.9333	0.7529
13	12	1	0.9167	0.6902
16	11	1	0.9091	0.6275
22	7	1	0.8571	0.5378
23	6	1	0.8333	0.4482

Any term in the product which has $d_j = 0$ can be omitted without affecting (4.4). The estimate $\hat{\mathscr{F}}(t)$ is, therefore, formally independent of the selection of points a_j for which the observed number of failures is zero. Thus, $\hat{\mathscr{F}}(t)$ is a function of the data only. It can, in fact, be shown to maximize the (generalized) likelihood over the space of all distributions, although this property has only limited direct statistical implications. Usually, $\hat{\mathscr{F}}(t)$ is called the Kaplan–Meier, or product-limit estimator. Table 4.1 shows an example of its calculation, for the leukaemia data of Table 1.1.

4.3 Greenwood's formula

If the possible failure times $a_1, a_2, \ldots, a_g$ are fixed, and the censoring mechanism allows the numbers of failures d_j at each a_j to increase at the same rate as the total sample size n, then the standard large-sample theory for maximum likelihood estimators applies, and the methods outlined in Chapter 3 may be used to make inferences about the $\hat{h}_j$ or functions of them such as $\hat{\mathscr{F}}(t)$.

Thus, asymptotically, $\sqrt{n}(\hat{h}_j - h_j)$ will have a multivariate normal distribution with mean zero and a covariance matrix which can be estimated by the inverse of the observed information matrix. Here

$$\left[\frac{\partial^2 l}{\partial h_j \partial h_k}\right]_{\hat{h}} = \begin{cases} -\dfrac{r_j}{\hat{h}_j(1 - \hat{h}_j)} & (j = k) \\ 0 & (j \neq k), \end{cases}$$

the same as would be obtained for k independent binomials. Since

$$\log \hat{\mathscr{F}}(t) = \sum^{(t)} \log(1 - \hat{h}_j),$$

and we have just seen that the $\hat{h}_j$ are asymptotically independent, the asymptotic variance of $\log \hat{\mathscr{F}}(t)$ and hence of $\hat{\mathscr{F}}(t)$ can easily be found, for any fixed t. Thus

$$\text{var}[\log \hat{\mathscr{F}}(t)] \approx \sum^{(t)} \text{var}[\log(1 - \hat{h}_j)]$$

$$\approx \sum^{(t)} \left(\frac{1}{1 - \hat{h}_j}\right)^2 \text{var}(\hat{h}_j)$$

$$\approx \sum^{(t)} \left(\frac{1}{1 - \hat{h}_j}\right)^2 \frac{\hat{h}_j(1 - \hat{h}_j)}{r_j}$$

$$= \sum^{(t)} \frac{d_j}{r_j(r_j - d_j)},$$

and

$$\text{var}[\hat{\mathscr{F}}(t)] = [\hat{\mathscr{F}}(t)]^2 \sum^{(t)} \frac{d_j}{r_j(r_j - d_j)}. \tag{4.5}$$

This is known as Greenwood's formula.

Confidence limits can now be obtained via the normal approximation based either on $\hat{\mathscr{F}}(t)$ or on $\log \hat{\mathscr{F}}(t)$. Limits such as

$$\hat{\mathscr{F}}(t_0) \pm k_\alpha^* \{\text{var}[\hat{\mathscr{F}}(t_0)]\}^{1/2},$$

where $\Phi(-k_\alpha^*) = \alpha$, refer to a prespecified t_0. A larger multiplier would be needed for a simultaneous confidence band for the function $\mathscr{F}(t)$ over some interval.

Some authors have suggested that (4.5) may be unstable in the tail of the distribution and have proposed an alternative, simpler estimate, namely,

$$\text{var}[\hat{\mathscr{F}}(t)] = \frac{[\hat{\mathscr{F}}(t)]^2 [1 - \hat{\mathscr{F}}(t)]}{r(t)}. \tag{4.6}$$

A rationale for (4.6) is as follows. Given the values of $n, r(t)$ and $\hat{\mathscr{F}}(t)$, it is plausible that the least informative configuration of the data is when all the censoring in $(0, t)$ occurs at the origin, so that the censored observations contribute no information to the estimation of $\mathscr{F}(t)$. In that case, the number of uncensored observations would be $r(t)/\hat{\mathscr{F}}(t) = n_0$, and (4.6) is obtained as the variance of a single binomial proportion $\hat{\mathscr{F}}(t)$ based on n_0 trials.

As explained in Section 3.3, the dependence on the function of $\mathscr{F}(t_0)$ chosen as the basis of the normal approximation can be avoided by the use of likelihood based confidence intervals. These can be derived as follows, working from the binomial log likelihood (4.3). The maximized log likelihood when the h_j are unconstrained in the unit cube is

$$\sum \{d_j \log(d_j/r_j) + (r_j - d_j) \log[(r_j - d_j)/r_j]\}.$$

Now suppose that $\theta = \mathscr{F}(t)$ is regarded as the parameter of interest:

$$\sum^{(t)} \log(1 - h_j) = \log \theta.$$

To test the null hypothesis $\theta = \theta_0$, we introduce a Lagrange multiplier ζ_0 and maximize instead of (4.3)

$$\sum d_j \log h_j + \sum (r_j - d_j) \log(1 - h_j) + \sum^{(t)} \zeta_0 \log(1 - h_j).$$

The maximizing values are $\tilde{h}_j$, with

$$\tilde{h}_j = \hat{h}_j \qquad (a_j \geqslant t),$$
$$\tilde{h}_j = d_j/(r_j + \zeta_0) \qquad (a_j < t),$$

where ζ_0 is determined by

$$\sum{}^{(t)} \log(1 - \tilde{h}_j) = \log\theta_0. \qquad (4.7)$$

The statistic for testing the null hypothesis is

$$W(\theta_0) = 2\sum{}^{(t)}\{r_j\log[(r_j + \zeta_0)/r_j]$$
$$+ (r_j - d_j)\log[(r_j - d_j)/(r_j + \zeta_0 - d_j)]\}. \qquad (4.8)$$

A $1 - \alpha$ confidence region for $\theta = \mathscr{F}(t)$ is now formed by taking

$$\{\theta; W(\theta) \leqslant c^*_{1\alpha}\},$$

where $c^*_{1\alpha}$ is the upper α point of the chi-squared distribution with one degree of freedom. This is best achieved by taking ζ as a new (data-dependent) parameter, obtaining from W confidence regions for ζ and then recalibrating the ζ scale in terms of θ by (4.7).

The relation with Greenwood's formula is made explicit by the expansions

$$W(\theta_0) = \zeta_0^2\sum{}^{(t)}\frac{d_j}{r_j(r_j - d_j)} + O_p(1/\sqrt{n})$$

$$= [\log\widehat{\mathscr{F}}(t_0) - \log\theta_0]^2\left(-\sum{}^{(t)}\frac{d_j}{r_j(r_j - d_j)}\right)^{-1} + O_p(1/\sqrt{n}).$$

The difference between confidence limits derived this way and those obtained from Greenwood's formula is most pronounced in the tails of the distribution, where asymmetric limits are most natural. Thus in the data analysed in Table 4.1, the first value of the survivor function, estimated as 0.8571, has a standard error from Greenwood's formula of 0.0764, and calculation of limits via a normal approximation is hazardous, and impossible at the more extreme levels. The values of ζ corresponding to $\alpha = 0.95$ are 59 and -11.9 and the resulting limits for $\mathscr{F}(t)$ are 0.9625 and 0.6703. At the final value recorded, with estimated survivor function 0.4482, with standard error 0.1346, the values of ζ are 12.3 and -3.75, with limits for $\mathscr{F}(t)$ of 0.6965 and 0.2028, almost symmetrical and close to the values based on Greenwood's formula.

4.4 Actuarial estimator

In practice, the distribution may be continuous rather than discrete, as has been assumed so far in this chapter. In this section, we consider the estimation of a continuous distribution with piecewise-constant hazard rate, that is for $j = 1, \ldots, g$,

$$h_T(t) = \frac{f(t)}{\mathscr{F}(t)} = \rho_j \qquad (a_{j-1} \leqslant t < a_j),$$

where the a_j are prespecified, with the convention that $a_0 = 0$. This representation is not especially plausible, but it does allow the estimated hazard rate to reflect the behaviour of the data in a way that is not possible under strong parametric assumptions, while avoiding the analytic complexities of fully nonparametric estimation of continuous distributions. The use of splines would allow additional smoothness conditions to be introduced into $h(t)$, at some cost in computational complexity, and with the disadvantage of possibly allowing $h(t)$ to be negative over part of its range.

We shall assume that censoring is also governed by a random mechanism, with its own piecewise-constant hazard

$$h_c(t) = \lambda_j \qquad (a_{j-1} \leqslant t < a_j),$$

and we set $b_j = a_j - a_{j-1}$, the interval width. We consider maximum likelihood estimation of the parameters ρ_j, the λ_j being regarded as nuisance parameters, first when all the survival times and censoring times are reported exactly, and secondly when they are given in grouped form. The latter case, where only the numbers d_j of failures and m_j of censorings in each interval $[a_{j-1}, a_j)$ are recorded, out of the r_{j-1} subjects entering that interval, is the more commonly encountered.

In the first case, the log likelihood in the ρ_j may be derived without reference to the λ_j by the methods of Chapter 3. In fact, from (3.10),

$$l(\rho) = \sum_{j=1}^{g} (d_j \log \rho_j - u_j \rho_j),$$

where

$$u_j = \sum_{i=1}^{n} I(x_i; a_{j-1}, a_j),$$

is the total time at risk in the interval (a_{j-1}, a_j). Here the function I is defined as

$$I(x; a_{j-1}, a_j) = \begin{cases} 0 & (x < a_{j-1}), \\ x - a_{j-1} & (a_{j-1} \leqslant x < a_j), \\ a_j & (x \geqslant a_j). \end{cases}$$

The maximum likelihood estimator of ρ_j, obtained by solving $\partial l/\partial \rho_j = 0$, is

$$\hat{\rho}_j = d_j/u_j, \tag{4.9}$$

the total number of failures in $[a_{j-1}, a_j)$ divided by the total time at risk in that interval, a straightforward extension of the result (3.13) for a single exponential parameter ρ. Since $\partial^2 l/\partial \rho_j \partial \rho_k \equiv 0 \; (j \neq k)$, the $\hat{\rho}_j$ are asymptotically independent.

In the second case, we must consider, separately for each interval $(a_{j-1}, a_j]$, the contribution to the joint likelihood $\mathrm{lik}(\rho, \lambda)$:

(i) from the $r_{j-1} - d_j - m_j$ subjects who survive uncensored throughout the interval;
(ii) from the d_j subjects who fail during the interval;
(iii) from the m_j subjects censored during the interval;

all conditioned on survival to the start of the interval. For clarity, we temporarily drop the subscripts $j-1$ and j. The conditional probabilities for the three events are

(i) $\exp[-b(\rho+\lambda)]$,

(ii) $\displaystyle\int_0^b \rho e^{-\rho v} e^{-\lambda v} dv = \frac{\rho}{\rho + \lambda}\{1 - \exp[-b(\rho + \lambda)]\}$,

(iii) $\displaystyle\int_0^b \lambda e^{-\lambda v} e^{-\rho v} dv = \frac{\lambda}{\rho + \lambda}\{1 - \exp[-b(\rho + \lambda)]\}$.

The contribution to the total log likelihood arising from the interval $(a_{j-1}, a_j]$ is

$$l_j(\rho_j, \lambda_j) = -(r - d - m)b(\rho + \lambda) + d\log\left(\frac{\rho}{\rho + \lambda}\right)$$
$$+ m\log\left(\frac{\lambda}{\rho + \lambda}\right) + (d + m)\log\{1 - \exp[-b(\rho + \lambda)]\},$$

where again the subscripts $j-1$ and j have been dropped from the right-hand side of the equation. As before, no other interval contributes to the log likelihood in (ρ_j, λ_j).

The maximum likelihood estimates $\hat{\rho}_j$ and $\hat{\lambda}_j$ can be obtained explicitly as the solutions of

$$\frac{\partial l}{\partial \rho} = -(r - d - m)b + \frac{d}{\rho} - \frac{d + m}{\rho + \lambda} + \frac{(d + m)b \exp[-b(\rho + \lambda)]}{1 - \exp[-b(\rho + \lambda)]} = 0,$$

$$\frac{\partial l}{\partial \lambda} = -(r - d - m)b + \frac{m}{\lambda} - \frac{d + m}{\rho + \lambda} + \frac{(d + m)b \exp[-b(\rho + \lambda)]}{1 - \exp[-b(\rho + \lambda)]} = 0,$$

$$(4.10)$$

for subtraction of the two equations gives

$$d/\rho = m/\lambda = (d + m)/(\rho + \lambda),$$

and substitution gives

$$\hat{\rho} = -\frac{d}{b(d + m)} \log\left(\frac{r - d - m}{r}\right), \qquad \hat{\lambda} = -\frac{m}{b(d + m)} \log\left(\frac{r - d - m}{r}\right).$$

$$(4.11)$$

If the interval width b is small, then $(d + m)/r$ will also be small and the logarithm may be expanded in series, giving

$$b\hat{\rho} = \frac{d}{r} + \frac{1}{2}\frac{d(d + m)}{r^2} + O\left(\frac{d + m}{r}\right)^3$$

$$= \frac{d}{r - \frac{1}{2}(d + m)}\left[1 + O\left(\frac{d + m}{r}\right)^2\right].$$

The estimator

$$\hat{\rho}_j = \frac{d_j}{b_j[r_{j-1} - \frac{1}{2}(d_j + m_j)]} \qquad (4.12)$$

is traditionally used to estimate the hazard rate ρ_j in the interval $[a_{j-1}, a_j)$. Comparison with (4.9) shows that the use of this estimator is in a sense equivalent to assuming that the deaths and censorings occur uniformly throughout the interval, when the denominator of (4.12) would equal u_j, the total time at risk in the interval.

Of generally greater interest than ρ_j itself, is the conditional probability $\exp(-b_j \rho_j)$ of survival throughout the interval in the absence of censoring. To the same order of approximation, the estimated probability of failure during the interval is

$$1 - \exp(-b_j \rho_j) \approx d_j/r'_j = \tilde{q}_j,$$

say, where r'_j, called the adjusted number at risk in $[a_{j-1}, a_j)$, is given by

$$r'_j = r_{j-1} - \tfrac{1}{2}m_j.$$

The actuarial estimator of $\mathscr{F}$ is obtained by combining the $\bar{q}_j$. It has the form

$$\tilde{\mathscr{F}}(a_j) = \prod_{k \leqslant j}\left(1 - \frac{d_k}{r'_k}\right). \tag{4.13}$$

This estimator differs from the product-limit estimator of Section 4.2 evaluated at $t = a_j+$ only in the 'half-period' correction of replacing r_j by r'_j. Usually the two estimates differ little, unless the data are heavily tied. Greenwood's formula for the estimated variance of $\tilde{\mathscr{F}}(a_j)$ becomes

$$\mathrm{var}[\tilde{\mathscr{F}}(a_j)] = [\tilde{\mathscr{F}}(a_j)]^2 \sum_{k=1}^{j} \frac{-d_k}{r'_k(r'_k - d_k)}.$$

4.5 Cumulative hazard estimators: goodness of fit

As mentioned in Section 2.4, plots of the hazard or cumulative hazard function are often useful in assessing the fit of a parametric family of survival distributions to a given set of data. Although minus the logarithm of the Kaplan–Meier estimator could be used to estimate the cumulative hazard, it is more usual to take

$$\hat{H}(t) = \sum\nolimits^{(t)} d_j/r_j, \tag{4.14}$$

as suggested by equation (2.12). Notice that if there are no ties and no censoring, so that $\{a_1, a_2, \ldots, a_g\}$ denote the ordered failure times, then

$$\hat{H}(a_k) = e_{nk} = \sum_{j=1}^{k} \frac{1}{n+1-j},$$

the expected value of the kth-order statistic in a unit exponential sample.

The estimated variance of $\hat{H}(t) = -\log\tilde{\mathscr{F}}(t)$ was obtained en route to Greenwood's formula as

$$\mathrm{var}[\hat{H}(t)] = \sum\nolimits^{(t)} \frac{d_j}{r_j(r_j - d_j)}. \tag{4.15}$$

Cumulative hazard plots have the disadvantage of tending to place too much visual emphasis on the behaviour in the tail of the distribution, where, as is clear from (4.15), the estimate is most unstable. Moreover, the sampling errors in $\hat{H}(t)$ are not independent.

With small amounts of data, a visual appraisal of random error may be desired, by inspection of the variation among points with independent errors, preferably of constant known variance. One approximate way of achieving this is as follows. Choose a small integer k, for example $k = 4$. Let $S_1^{(k)}, S_2^{(k)}, \ldots$, be the total time at risk up to the kth failure, between the kth and $2k$th failure, etc. If ρ_j denotes the average total hazard during the period defining $S_j^{(k)}$, i.e. summed over all subjects at risk in that period, then $2\rho_j S_j^{(k)}$ is distributed as chi-squared with $2k$ degrees of freedom. It follows from standard properties of the log chi-squared distribution that

$$Z_j^{(k)} = -\log\left(\frac{S_j^{(k)}}{k}\right) - \frac{1}{2k - \frac{1}{3}}$$

has mean approximately $\log \rho_j$ and variance approximately $(k - \frac{1}{2})^{-1}$. Further, the $Z_j^{(k)}$ are approximately mutually independent. A natural plot is thus of $Z_j^{(k)}$ versus $\bar{t}_j^{(k)}$ or $\log \bar{t}_j^{(k)}$, where $\bar{t}_j^{(k)}$ is the time at the centre of the relevant interval.

4.6   Bayesian nonparametric methods

In the absence of censoring, the natural prior distribution for describing uncertainty about the atoms f_j of the discrete possibly improper survival distribution (2.6) is the Dirichlet distribution, with density proportional to

$$f_1^{\alpha_1 - 1} f_2^{\alpha_2 - 1} \ldots f_g^{\alpha_g - 1} (1 - f_1 - f_2 - \ldots - f_g)^{\alpha_{g+1} - 1}$$

(4.16)

for $\alpha_j \geqslant 0$ $(j = 1, 2, \ldots, g + 1)$. This is the conjugate prior for the multinomial likelihood, in the sense that the posterior distribution for $\{f_j\}$, given that, of a total of n failures, d_j are observed to occur at $a_j (\sum d_j \leqslant n)$, is of the same form with parameters $\alpha_j' = \alpha_j + d_j (j = 1, \ldots, g), \alpha_{g+1}' = \alpha_{g+1} + n - \sum d_j$.

When some of the failure times are censored, however, the posterior distribution of f is no longer of the Dirichlet form. As usual, it is more convenient to transform to the (discrete) hazard function, via the transformation (4.1). The Jacobian of this transformation is

$$(1 - h_1)^{g-1}(1 - h_2)^{g-2} \ldots (1 - h_{g-1})^1,$$

and, as is evident from the definitions of f and h,

$$1 - f_1 - f_2 - \ldots - f_g = (1 - h_1)(1 - h_2)\ldots(1 - h_g),$$

both expressions equalling $\text{pr}(T > a_g)$. Expressed in terms of $\{h_j\}$ the prior (4.16) becomes, apart from a proportionality constant,

$$\prod_{j=1}^{g} h_j^{\alpha_j - 1}(1 - h_j)^{\gamma_j - 1}, \qquad (4.17)$$

where

$$\gamma_j = \sum_{k=j+1}^{g+1} \alpha_k. \qquad (4.18)$$

The expression (4.17) has the form of a product of independent beta distributions. The degenerate form of (4.16), with $\alpha_j = 0$ for all j, corresponds to the degenerate form of (4.17), with $\alpha_j = \gamma_j = 0$, for all j. In general, however, exchangeability of the f_j, i.e. $\alpha_1 = \alpha_2 = \ldots = \alpha_g$, is not equivalent to exchangeability of the h_j.

As noted earlier, the likelihood for $\{h_j\}$, from right-censored data in which d_j failures are observed among the r_j subjects at risk at a_j, is proportional to

$$\prod_{j=1}^{g} h_j^{d_j}(1 - h_j)^{r_j - d_j}, \qquad (4.19)$$

equivalent to that from a product of independent binomials. From (4.17) and (4.19) the posterior distribution of $\{h_j\}$, given the data, is also a product of independent beta distributions, with indices $\alpha_j' = \alpha_j + d_j$, $\gamma_j' = \gamma_j + r_j - d_j$. This will not in general correspond to a Dirichlet distribution over the $\{f_j\}$.

The posterior distribution of the survivor function $\mathscr{F}(a_j) = (1 - h_1)(1 - h_2)\ldots(1 - h_{j-1})$ can in theory be derived, but it does not take a simple form. The posterior moments of $\mathscr{F}(a_j)$ are easily obtained (Exercise 4.7). For the degenerate 'ignorance' prior with $\alpha_j = \gamma_j = 0$, the product-limit estimator is recovered as the posterior mean of $\mathscr{F}$.

Bibliographic notes, 4

Life tables have been used by demographers and actuaries for many years to describe and compare patterns of human mortality, often via the so-called expectation of life. The product-limit estimator appears first to have been proposed by Böhmer (1912) but the actuarial estimator itself is much older. Formula (4.5) is due to Greenwood

(1926). Kaplan and Meier (1958) derived the product-limit estimator from maximum likelihood arguments; for further discussion see Johansen (1978). A key reference is Efron (1967); see Chapter 11 of this book. Anderson and Senthilselvan (1980) discuss the use of splines in the estimation of hazard functions.

At the price of some rather artificial restrictions on the form of the survival distribution, we have been able to appeal to standard large-sample likelihood theory to justify the methods. These restrictions can largely be dispensed with at the cost of increasing the complexity of the theory. The notion of weak convergence (Billingsley, 1968, 1971) provides the appropriate mathematical tool. For applications to censored data, see Breslow and Crowley (1974) and Meier (1975). Gillespie and Fisher (1979) and Hall and Wellner (1980) use the theory to derive simultaneous confidence bands. See Aalen (1976, 1978) and Gill (1980) for an approach using martingale theory. Reid (1981a) discusses influence functions for censored data, and Efron (1981) and Reid (1981b) describe some numerical methods for obtaining interval estimates of the median of a survival distribution from censored data.

A comment similar to that of the preceding paragraph applies to the Bayesian theory. Ferguson (1973) introduced Dirichlet processes, for the purpose of deriving nonparametric Bayesian estimates of a distribution function from uncensored data. This avoids the need to specify the atoms of the distribution in advance. Extensions to censored survival times are discussed by Susarla and Van Ryzin (1976), Doksum (1974), Cornfield and Detre (1977), Kalbfleisch and Mackay (1978), Kalbfleisch (1978) and Ferguson and Phadia (1979). Burridge (1981b) describes an empirical Bayes approach.

The cumulative hazard plot is due to Nelson (1969, 1972). See Cox (1979) for discussion of graphical methods for assessment of fit.

Further results and exercises, 4

4.1. Verify that in the absence of censoring, the product-limit estimator reduces to the empirical distribution function

$$\hat{F}(t) = \frac{1}{n} \sum_{i=1}^{n} I(x_i, t),$$

where $I(x, t) = 1$ if $x < t$, $I(x, t) = 0$ if $x \geqslant t$. Show also that Greenwood's formula reduces to the usual binomial variance estimate in this case.

4.2. Suppose that the censoring times are random with survivor function $K(t)$. Let $Z(t) = \sqrt{n}[\hat{\mathscr{F}}_n(t) - \mathscr{F}(t)]$, where $\hat{\mathscr{F}}_n(t)$ is the product-limit estimator of the survivor function based on n observations. Show informally that as $n \to \infty$, the covariance function $\text{cov}[Z_n(s), Z_n(t)]$ $(s \leqslant t)$ converges to

$$- \mathscr{F}_0(s)\mathscr{F}_0(t) \int_{u=0}^{s} \frac{d\mathscr{F}(u)}{[\mathscr{F}(u)]^2 K(u)}.$$

[For a rigorous proof, see Breslow and Crowley, 1974.]

4.3. The restricted mean of F is estimated from the Kaplan–Meier estimator as

$$\hat{\mu}_L = \int_0^L \hat{\mathscr{F}}(t)\,dt.$$

Show from the preceding exercise that the variance of $\hat{\mu}_L$ may be estimated by

$$\text{var}(\hat{\mu}_L) = \frac{1}{n} \int_0^L \frac{1}{[\mathscr{F}(s)]^2 K(s)} \left(\int_s^L \mathscr{F}(u)\,du \right)^2 |d\mathscr{F}(s)|.$$

What happens as $L \to \infty$? Discuss how the variance of $\hat{\mu}_L$ and of its limit (the so-called expectation of life) may be estimated.
[Irwin, 1949; Kaplan and Meier, 1958; Meier, 1975; Reid, 1981a]

4.4. In Section 4.4, show that the exact conditional distribution of d_j given r_{j-1} and $d_j + m_j$, is binomial, with index $d_j + m_j$ and parameter $\rho/(\rho + \lambda)$.

4.5. Examine the extent to which the arguments of Section 4.4 hold if it is assumed that the hazard functions $h_T(t)$ and $h_c(t)$ in (a_{j-1}, a_j) take the form

$$h_T(t) = \rho_j h_0(t), \qquad h_c(t) = \lambda_j h_0(t),$$

for some unknown function $h_0(t)$.

4.6. Compare the second derivatives of the log likelihoods for the two cases of Section 4.4. Find an expression for the expected loss of information about ρ_j through the grouping, and evaluate this for representative values of $b\rho_j, b\lambda_j$.

[Pierce *et al.*, 1979]

4.7. Show that if h has a beta distribution, with density proportional to $h^{\alpha-1}(1-h)^{\gamma-1}$, then its expected value is $\alpha/(\alpha+\gamma)$ if $\alpha>0$, $\gamma>0$, zero if $\alpha=0$, $\gamma>0$, and unity if $\alpha>0$, $\gamma=0$. Hence derive the posterior mean estimate of the survivor function in Section 4.6.

4.8. Prove via the argument used to derive Greenwood's formula that the approximate variance of the logit transform, $\log\{\hat{\mathscr{F}}(t)/[1-\hat{\mathscr{F}}(t)]\}$, is

$$[1-\hat{\mathscr{F}}(t)]^{-2}\sum^{(t)}\frac{d_j}{r_j(r_j-d_j)}.$$

Suggest how this could be used for the analysis by empirically weighted least squares of a linear logistic model for the survivor function at a single specified point t_0, the data being grouped into sets with equal or nearly equal values of the explanatory variables.

4.9. A smoothed estimate of the hazard function can be produced from uncensored data by finding a 'kernel estimate' of the density and

(i)   dividing by the sample survivor function;
(ii) dividing by the integral of the smoothed density or by some other smoothed survivor function.

Compare these and comment on the advisability of a kernel that is different in different parts of the range.

How would these procedures and discussion be adapted to censored data? Under what circumstances is it desirable to use graphical and numerical procedures capable of producing smooth answers out of the most limited data as contrasted with procedures in which the intrinsic variability is shown explicitly?

CHAPTER 5

Dependence on explanatory variables: model formulation

5.1 Introduction

In the previous chapters we have discussed models and analysis for relatively simple problems involving a single distribution. When, as is often the case, two or more sets of data have to be compared, this is sometimes best done by estimating survivor functions for each set of data separately and then making a qualitative comparison, either directly or via summary statistics. More sensitive or more complicated comparisons are, however, best handled by comprehensive models in which the effect of the explanatory variables is typically represented by unknown parameters.

In the present chapter, we review some of the many possible models that may be used to represent the effect on failure time of explanatory variables. For this we suppose that for each individual there is defined a $q \times 1$ vector z of explanatory variables. The components of z may represent various features thought to affect failure time, such as

  (i)   treatments;
  (ii)   intrinsic properties of the individuals;
  (iii) exogenous variables.

Further components of z may be synthesized to examine interaction effects, in a way that is broadly familiar from multiple regression analysis.

The explanatory variables may be classified also in other ways, in particular as for each individual constant or time-dependent. Some of the ideas that follow do not apply to time-dependent explanatory variables; further, for many of the statistical techniques, computation is much harder for time-dependent explanatory variables. Nevertheless, for a variety of reasons that will appear later, it is important to accommodate time-dependent explanatory variables in the discussion.

A few outline examples of explanatory variables are as follows. In a simple comparison of two treatments, for instance of a 'new' treatment with a 'control', we consider a binary explanatory variable equal to one for individuals receiving the treatment, and equal to zero for those receiving the control. If the treatment is specified by a dose (or stress) level, the corresponding explanatory variable is dose or log dose. If the treatment is factorial, several explanatory variables will be required with synthesized product variables to represent interactions where appropriate. In most cases such variables will be constant for each individual, but time-dependent treatment variables can arise in two rather different ways.

First, especially in some industrial reliability contexts, a time-varying stress may be applied, or some cumulative measure of total load may be judged relevant. In defining the components of the vector z, it may then be convenient to introduce functions of the whole previous history of the dose or stress process.

Secondly, it may happen that a treatment under study is not applied until some time after the time origin. Then a suitable explanatory variable may be a time-dependent binary variable that jumps from 0 to 1 at the point of application of the treatment.

Explanatory variables measuring intrinsic properties of individuals include, in a medical context, such demographic variables as sex, age on entry, and variables describing medical history before admission to the study. Other variables may define qualitative groupings of the individuals.

Considerable care is needed in introducing as explanatory variables time-dependent variables that may be influenced by the treatment variables under investigation. For example, consider the comparison of the effect on survival of two alternative treatments for the control of hypertension, blood pressure being an explanatory variable. The use of blood pressure before treatment assignment as a fixed explanatory variable is a standard device for precision improvement and interaction detection. The inclusion of blood pressure monitored after treatment assignment as an explanatory variable would address the question as to whether any difference in survival between the treatments is explained by the effect on blood pressure control. Time-dependent explanatory variables are discussed in a little more detail in Chapter 8.

Finally exogenous variables define, in particular, environmental features of the problem and may be needed also to represent groupings of the individuals corresponding to observers, sets of apparatus, etc.

It is often convenient to define the vector z of explanatory variables so that $z = 0$ corresponds to some meaningful 'standard' set of conditions, for example a control treatment. Frequently models can conveniently be developed in two parts:

(a) a model for the distribution of failure time when $z = 0$;
(b) a representation of the change induced by a nonzero z, often in terms of some parametric form.

In the description of particular models that follows, it is often convenient to start with the simplest case of the comparison of two treatments, corresponding to a single binary explanatory variable, the generalization usually being obvious. Throughout $\psi(z)$ denotes a function linking z to survival: increasing $\psi(z)$ always corresponds to increasing risk, i.e. to decreasing failure time. The symbol β is reserved for a parameter vector characterizing $\psi(z)$. Note that the functions $\psi(z)$ in two different models are not in general quantitatively comparable.

5.2   Accelerated life model

(i) Simple form

Suppose that there are two treatments represented by values 0 and 1 of the explanatory variable z. Let the survivor function at $z = 0$ be $\mathcal{F}_0(t)$; in the accelerated life model there is a constant ψ such that the survivor function at $z = 1$, written variously $\mathcal{F}_1(t)$ or $\mathcal{F}(t; 1)$, is

$$\mathcal{F}_1(t) = \mathcal{F}_0(\psi t), \tag{5.1}$$

so that

$$f_1(t) = \psi f_0(\psi t), \qquad h_1(t) = \psi h_0(\psi t). \tag{5.2}$$

A stronger version is that any individual having survival time t under $z = 0$ would have survival time t/ψ under $z = 1$, i.e. the corresponding random variables are related by $T_1 = T_0/\psi$.

More generally, with an arbitrary constant vector z of explanatory variables, suppose that there is a function $\psi(z)$ such that the survivor function, density and hazard are respectively

$$\mathcal{F}(t;z) = \mathcal{F}_0[t\psi(z)],$$
$$f(t;z) = f_0[t\psi(z)]\psi(z),$$
$$h(t;z) = h_0[t\psi(z)]\psi(z).$$

(5.3)

If $\mathcal{F}_0(\,.\,)$ refers to the standard conditions $z = 0$, then $\psi(0) = 1$. A representation in terms of random variables is

$$T = T_0/\psi(z),$$

(5.4)

where T_0 has survivor function $\mathcal{F}_0(\,.\,)$. If $\mu_0 = E(\log T_0)$, we can write this as

$$\log T = \mu_0 - \log \psi(z) + \varepsilon,$$

(5.5)

where ε is a random variable of zero mean and with a distribution not depending on z.

In problems with a limited number of distinct values of z, it may be unnecessary to specify $\psi(\,.\,)$ further. In other contexts, a parametric form for $\psi(\,.\,)$ may be needed; we then write $\psi(z;\beta)$. Since $\psi(z;\beta) \geqslant 0$, $\psi(0;\beta) = 1$, a natural candidate is

$$\psi(z;\beta) = e^{\beta^T z},$$

(5.6)

where now the parameter vector β is $q \times 1$. Then (5.5) can be written

$$\log T = \mu_0 - \beta^T z + \varepsilon,$$

(5.7)

a linear regression model. Note that for the comparison of two groups, with a single binary explanatory variable, we get (5.1) with $\psi = e^\beta$.

(ii) Some consequences useful for model checking

The central property of the accelerated life model can be re-expressed in various ways that can be used as a basis for testing the adequacy of the model. Thus from (5.5) the distributions of $\log T$ at various values of z differ only by translation. In particular $\text{var}(\log T)$ is constant.

Alternatively in the two-sample problem we can compare quantiles. We define $t_0^{(a)}, t_1^{(a)}$, for $0 < a < 1$, by

$$a = \mathcal{F}_0(t_0^{(a)}) \qquad t_0^{(a)} = \mathcal{F}_0^{-1}(a),$$
$$a = \mathcal{F}_1(t_1^{(a)}), \qquad t_1^{(a)} = \mathcal{F}_1^{-1}(a),$$

(5.8)

Table 5.1 *Spring data of Table 1.3; T is thousands of cycles to failure. Mean and standard deviation of log T*

Stress (N/mm^2)	Log T	
	Mean	SD
950	5.11	0.214
900	5.35	0.198
850	5.84	0.172
800	6.52	0.580
750	8.41	0.735

so that under (5.1)

$$t_1^{(a)} = t_0^{(a)}/\psi,$$

i.e. the so-called Q–Q plots (quantile–quantile plots) are straight lines through the origin. We have for simplicity assumed in (5.8) that $\mathscr{F}_0(.)$ is strictly decreasing, so that the quantiles are uniquely defined.

Quite often a simple analysis will show the accelerated life model to be inadequate. For instance, with the spring data of Table 1.3, inspection suggests that the relative dispersion increases substantially as stress decreases and mean failure time increases. This is confirmed by calculating the mean and standard deviation of log T at each stress level. For this the lowest stress level at which the censoring is severe has been omitted and for the next lowest stress the single censored value has been treated as a failure, in this instance tending to underestimate the mean and standard deviation of log T. The results are summarized in Table 5.1.

The accelerated life model may hold at the three highest stress levels, but there is overwhelming evidence against it as a representation for the whole stress range. Further interpretation would be aided by information on the mode of failure, in particular as to whether this is different at the lowest stress levels. At 750 N/mm^2 the distribution is close to an exponential with a nonzero starting value.

(iii) Time-dependent explanatory variables

Suppose now that the explanatory variable z is time-dependent, $\{z(t)\}$, say. First, it will usually be good to define $z(t)$ so that the hazard at any particular time t depends only on the explanatory variable at that time. This may involve introducing as components of $z(t)$ integrals, sums, derivatives and differences of the explanatory variables as originally recorded.

The essence of the accelerated life model is that 'time' is contracted or expanded relative to that at $z = 0$. This suggests that for an individual characterized by $z(t)$, time $t^{(z)}$, say, evolves relative to the time $t^{(0)}$ for that individual had he been at $z = 0$ in accordance with

$$dt^{(z)}/dt^{(0)} = 1/\psi[z(t^{(z)})],$$

i.e.
$$t^{(0)} = \int_0^{t^{(z)}} \psi[z(u)]\,du = \Psi(t^{(z)}), \tag{5.9}$$

say, so that the failure times are related, instead of by (5.4), by

$$T = \Psi^{-1}(T_0).$$

Note, however, that the result of applying two such transformations to T_0 will in general depend on the order in which they are applied, so that linear combinations of such time-dependent explanatory variables will not obey the commutativity relations of ordinary arithmetic.

Hence survivor function, density and hazard are

$$\mathscr{F}[t;\{z(.)\}] = \mathscr{F}_0[\Psi(t)],$$
$$f[t;\{z(.)\}] = \psi[z(t)]f_0[\Psi(t)], \tag{5.10}$$
$$h[t;\{z(.)\}] = \psi[z(t)]h_0[\Psi(t)].$$

(iv) Generality of the time-dependent model

The accelerated life model with time-varying explanatory variables has rarely been used in applications, so far as we know, although it would be appropriate for systems subject to nonconstant treatment variables, e.g. 'stress'. There is another sense, however, in which the use of time-varying explanatory variables converts the very special model into a very general one. Consider for simplicity the comparison of two groups and suppose that instead of a simple binary explanatory variable we introduce

$$z = \begin{cases} 0 & \text{group 0,} \\ \xi(t) & \text{group 1,} \end{cases}$$

where $\xi(t)$ is a function to be chosen and we take $\psi(z) = e^z$. Then by

(5.10) the survivor function in group 1 is

$$\mathscr{F}_0[\Lambda(t)],$$

where

$$\Lambda(t) = \int_0^t e^{\xi(u)}\, du.$$

Thus a given survivor function $\mathscr{F}_1(t)$ is reproduced by taking

$$e^{\xi(t)} = \frac{d}{dt}\mathscr{F}_0^{-1}\mathscr{F}_1(t), \tag{5.11}$$

it being assumed that the support of $\mathscr{F}_0(\,.\,)$ contains that of $\mathscr{F}_1$.

One way of producing a fairly rich family of models for the two-group problem is thus to write for $j = 0, \ldots, p$

$$z_j = \begin{cases} 0 & \text{group } 0, \\ t^j & \text{group } 1, \end{cases} \tag{5.12}$$

for some suitable value of p and then to take

$$\psi(z) = e^{\beta^{\mathrm{T}} z}, \tag{5.13}$$

where β is a $q \times 1$ parameter vector, $q = p + 1$.

In most instances this extension of the accelerated life model is a formal one without direct physical significance. Note that functions other than powers of t could be used in (5.12) and that the argument extends in principle to problems more complex than the comparison of two groups.

(v)   *Several types of failure*

One possible explanation of inconsistency with the accelerated life model is the presence of several types of failure, each following an accelerated life model but with different modifying functions ψ. As z varies, the balance between the types of failure changes. Of course, if the types of failure are observed, we can construct a more detailed model; see Chapter 9. If the distinct types of failure are not observable, it may sometimes be fruitful to hypothesize a small number of failure types, to attempt to deduce something about their properties by examining simple models and then to aim at further data to see whether the hypothesized failure types have physical identity.

Suppose then that there are l failure types, and that the observed failure time T can be represented as

$$T = \min(T_1, \ldots, T_l), \qquad (5.14)$$

where at $z = 0$ the T_j are independent random variables with survivor functions $\mathscr{F}_{0j}(\,.\,)$, possibly improper. Consider for simplicity the case of a single binary variable z and suppose that at $z = 1$ the survivor function of T_j is $\mathscr{F}_{0j}(\psi_j t)$. Then

$$\mathscr{F}_0(t) = \prod_j \mathscr{F}_{0j}(t), \qquad \mathscr{F}_1(t) = \prod_j \mathscr{F}_{0j}(\psi_j t)$$

and it follows easily that

$$h_0(t) = \sum_j h_{0j}(t), \qquad h_1(t) = \sum_j \psi_j h_{0j}(\psi_j t). \qquad (5.15)$$

(vi) Parametric version

So far the survivor function $\mathscr{F}_0(\,.\,)$ at $z = 0$ has been unspecified. If now we take $\mathscr{F}_0(\,.\,)$ to be a member of any of the parametric families discussed in Chapter 2, we obtain a special family of accelerated life models. If, further, $\psi(\,.\,)$ is specified parametrically, we have a fully parametric model. In particular, if the survivor function $\mathscr{F}_0(\,.\,)$ is log normal and $\psi(z; \beta) = e^{\beta^T z}$, the linear model (5.7) for $\log T$ is a normal-theory one and all the usual least-squares methods are available, provided that there is no censoring.

One important special case arises when $\mathscr{F}_0(\,.\,)$ is a Weibull distribution, (2.14), with parameters (ρ_0, κ), say. Then with constant explanatory variables, it is clear that T, for specified z, also has a Weibull distribution, with parameters $(\rho_0 \psi(z), \kappa)$. A special case of this is the exponential distribution, $\kappa = 1$.

The most important special case, however, is probably the log logistic, which is introduced by a rather different route in the next subsection.

(vii) Log logistic accelerated life model

If attention is concentrated on a particular time t_0, failure or nonfailure by time t_0 can be treated as a binary response. It is then natural to consider a linear logistic model in which

$$\log\{\mathscr{F}(t_0; z)/[1 - \mathscr{F}(t_0; z)]\} = \tilde{\beta}^T z + \alpha(t_0),$$

where $\alpha(t_0)$ refers to the baseline $z = 0$.

Now suppose that such a model is required to hold for all t_0. We could, of course, make $\tilde{\beta}$ as well as $\alpha(t)$ depend on t, but the simplest representation arises if $\tilde{\beta}$ is independent of t; we then require that $\alpha(t) \to \infty$ as $t \to 0$ and $\alpha(t) \to -\infty$ as $t \to \infty$. This can most simply be achieved by taking $\alpha(t)$ to be proportional to $-\log t$. If we write $\alpha(t) = -\kappa \log(t\rho)$ and $\tilde{\beta} = -\kappa\beta$, then

$$\mathscr{F}(t;z) = \frac{1}{1 + (\rho t e^{\beta^T z})^\kappa}. \qquad (5.16)$$

This is of precisely the accelerated life form with baseline survivor function

$$\frac{1}{1 + (t\rho)^\kappa},$$

the log logistic distribution; see Section 2.3(viii).

This representation can be extended in various ways as a basis for testing goodness of fit; one such, as noted above, is to allow β to depend on t.

5.3   Proportional hazards model

(i)   Simple form

A second broad family of models that has been widely used in the analysis of survival data is best specified via the hazard function. For a constant vector z of explanatory variables suppose that the hazard is

$$h(t;z) = \psi(z)h_0(t). \qquad (5.17)$$

Here $h_0(.)$ is the hazard for an individual under the standard conditions, $z = 0$, and we require $\psi(0) = 1$. The survivor function and density are thus

$$[\mathscr{F}_0(t)]^{\psi(z)}, \qquad \psi(z)[\mathscr{F}_0(t)]^{\psi(z)-1} f_0(t).$$

Thus the survivor functions form the Lehmann family generated from $\mathscr{F}_0(.)$. We call (5.17) the (simple) proportional hazards model.

Note that the function $\psi(z)$, while fulfilling the same role as the $\psi(z)$ of Section 5.2, does not have precisely the same interpretation. The function $\psi(z)$ can be parameterized, as in the previous discussion, as $\psi(z;\beta)$ and in particular the most important special case is again

$$\psi(z;\beta) = e^{\beta^T z}. \qquad (5.18)$$

The reasons for considering this model are that

(a) there is a simple easily understood interpretation to the idea that the effect of, say, a treatment is to multiply the hazard by a constant factor;
(b) there is in some fields empirical evidence to support the assumption of proportionality of hazards in distinct treatment groups;
(c) censoring and the occurrence of several types of failure are relatively easily accommodated within this formulation and in particular the technical problems of statistical inference when $h_0(t)$ is arbitrary have a simple solution.

(ii) Relation with accelerated life model

For constant explanatory variables the question naturally arises as to when the proportional hazards model (5.17) is also an accelerated life model. For this we need there to exist a function $\chi(z)$ such that

$$[\mathscr{F}_0(t)]^{\psi(z)} = \mathscr{F}_0[t\chi(z)].\tag{5.19}$$

Write

$$\mathscr{G}_0(\tau) = \log[-\log \mathscr{F}_0(e^\tau)].$$

Then

$$\log\psi(z) + \mathscr{G}_0(\tau) = \mathscr{G}_0[\tau + \lambda(z)],$$

where $\lambda(z) = \kappa^{-1}\log\chi(z)$. For this to hold for all τ and for some nonzero $\lambda(z)$, i.e. nonunit $\chi(z)$, we need

$$\mathscr{G}_0(\tau) = \kappa\tau + \alpha, \qquad \lambda(z) = \log\psi(z),$$

where α, κ are constants. Thus, on writing $\rho = e^{\alpha/\kappa}$, we have that

$$\mathscr{F}_0(t) = \exp[-(\rho t)^\kappa].$$

That is, the Weibull distribution is the only initial distribution for which, with constant explanatory variables, the accelerated life and proportional hazards models coincide.

It follows directly from the definition of the Weibull distribution that the accelerated life model with 'scale' parameters $\psi_{AL}(z)$ has survivor function and hazard

$$\exp\{-[\rho\psi_{AL}(z)t]^\kappa\}, \qquad \kappa[\rho\psi_{AL}(z)]^\kappa t^{\kappa-1},$$

i.e. is a proportional hazards model defined by

$$\psi_{PH}(z) = [\psi_{AL}(z)]^\kappa.$$

In particular if $\psi_{AL}(z) = \exp(\beta_{AL}^T z)$, then $\psi_{PH}(z) = \exp(\beta_{PH}^T z)$ with $\beta_{PH} = \kappa\beta_{AL}$.

The distinction between the proportional hazards and the accelerated life models is perhaps best seen from an artificial special case.

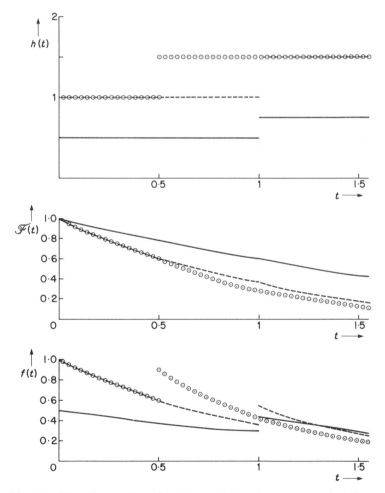

Fig. 5.1. Hazard, survivor function and density: ———, baseline; ○○○○, accelerated by a factor of 2; ---, with hazard doubled.

Fig. 5.1 shows a baseline hazard (and survivor function and density) of step function form and the corresponding functions first accelerated by a factor of 2 and then with the hazard multiplied by 2.

(iii) *Time-dependent explanatory variables*

The specification (5.17) of the simple proportional hazards model

$$h(t;z) = \psi(z)h_0(t)$$

extends immediately when the explanatory variable is time-dependent. Many of the remarks of Sections 5.2(iii) and (iv) concerning time-dependent variables and the accelerated life model are relevant here also. Thus we aim to define $z(t)$ so that $h(t;z)$ depends only on $z(t)$ and not on $z(t')$, $t' \neq t$. Also time-dependent $z(t)$ may be either values of relevant subject matter variables that change with time or may be derived variables included to test the applicability of or to generalize the special model (i). Thus in the comparison of two groups, the specification (5.12) and (5.13) gives hazards in the two groups respectively of

$$h_0(t), \qquad \exp\left(\sum_{j=0}^{p} \beta_j t^j\right) h_0(t) \qquad (5.20)$$

and so, at least for sufficiently large p, any two distributions with common support can be represented.

(iv) *Parametric version*

In the discussion so far, while the dependence on the explanatory variable has sometimes been parameterized, for instance in the form (5.18), the hazard at $z = 0, h_0(t)$, has been left arbitrary. Fully parametric models can be obtained by inserting for $h_0(t)$ the hazard for one of the families of distributions discussed in Chapter 2.

The most important special cases are probably the Weibull, including the exponential as a particular form, the Gompertz–Makeham and the log logistic.

5.4 Nonmultiplicative hazard-based model

In the simple proportional hazards model (5.17),

$$h(t; z) = \psi(z)h_0(t),$$

with constant explanatory variables, the hazard functions at different levels of z are proportional. It is possible, however, that the hazard functions are parallel rather than proportional or that, more generally, they can be made parallel by a nonlinear transformation of h. Thus we may consider the representation

$$h(t;z) = \phi(z) + h_0(t) \tag{5.21}$$

or

$$h^{(\lambda)}(t;z) = \phi(z) + h_0^{(\lambda)}(t), \tag{5.22}$$

where, for example,

$$h^{(\lambda)}(t;z) = \begin{cases} [h(t;z)]^\lambda & (\lambda \neq 0), \\ \log h(t;z) & (\lambda = 0). \end{cases} \tag{5.23}$$

In (5.21) and (5.22), $\phi(0) = 0$ and $\phi(z)$ is constrained so that the right-hand side is nonnegative. Parametric versions are obtained by, for instance, writing $\phi(z) = \beta^T z$ and by taking one of the parametric families of Chapter 2 for $h_0(t)$. With $\lambda = 0$ we recover the proportional hazards model.

The choice between these models will normally be an empirical matter, involving either the formal estimation of λ in (5.23) or, more commonly, the inspection of estimated hazard functions.

Some light is thrown on the alternative formulations by (5.14). That is, if

$$T = \min(T_1, \ldots, T_l), \tag{5.24}$$

where $T_1, \ldots, T_l$ are independent random variables, then

$$h(t) = \sum_{j=1}^l h_j(t), \tag{5.25}$$

where h_j is the hazard for T_j.

Now suppose that the difference between two treatments lay in the elimination of some of the T_j, i.e. some of the sources of failure. Then the difference between the two hazards would be

$$\sum{}' h_j(t),$$

where the sum is over the eliminated T_j. If the eliminated T_j were to have constant hazards over the range of interest, then an additive model (5.21) would result. If, on the other hand, the eliminated T_j were in some rough sense a large random sample from the l, then a proportional hazards model would result.

5.5 Transferred origin model

A possible relation between the hazard functions in two groups is that one is the translation in time of the other. That is, for some constant Δ the hazard functions are

$$h_0(t), \qquad h_0(t + \Delta),$$

with an obvious extension to $h_0[t + \Delta(z)]$ for the hazard corresponding to explanatory variable z.

Clearly $h_0(t + \Delta) = \psi h_0(t)$ for all t if and only if $h_0(t)$ is exponential, the Gompertz form, so that the transferred origin model is equivalent to the simple proportional hazards model only in this case. The model is directly meaningful only for values of t for which both t and $t + \Delta$ are positive.

5.6 Accelerated onset model

In some contexts the effect of a treatment may be to accelerate (or retard) the onset of failure in some individuals, leaving the remainder unchanged. Thus for accelerated onset an individual who has survived an appreciable time will have the same hazard function in the two groups.

One way of representing this is to postulate a mixture of two (or more) types of individuals with survivor functions $\mathscr{F}_{01}(t)$ and $\mathscr{F}_{02}(t)$, say. The distinction between the types is not directly observable. If the proportional hazards model applies to the first type, we have for the two survivor functions

$$
\begin{aligned}
\mathscr{F}_0(t) &= \theta \mathscr{F}_{01}(t) + (1 - \theta)\mathscr{F}_{02}(t), \\
\mathscr{F}_1(t) &= \theta [\mathscr{F}_{01}(t)]^{\psi} + (1 - \theta)\mathscr{F}_{02}(t),
\end{aligned}
\tag{5.26}
$$

the corresponding hazard functions being

$$
\begin{aligned}
h_0(t) &= [\theta f_{01}(t) + (1 - \theta)f_{02}(t)] \\
&\quad \times [\theta \mathscr{F}_{01}(t) + (1 - \theta)\mathscr{F}_{02}(t)]^{-1}, \\
h_1(t) &= \{\theta\psi f_{01}(t)[\mathscr{F}_{01}(t)]^{\psi - 1} + (1 - \theta)f_{02}(t)\} \\
&\quad \times \{\theta [\mathscr{F}_{01}(t)]^{\psi} + (1 - \theta)\mathscr{F}_{02}(t)\}^{-1}.
\end{aligned}
$$

In (5.26) we require $\mathscr{F}_{01}(t)$ and $[\mathscr{F}_{01}(t)]^{\psi}$ to converge to zero faster than $\mathscr{F}_{02}(t)$, as $t \to \infty$.

While under suitable circumstances such a model could be fitted following parameterization of the component survivor functions, it is likely that a large amount of high-quality data would be needed unless, perhaps, the component survivor functions are drastically restricted, for example by supposing one or both of them to be exponential. In the latter case the survivor functions are, say,

$$\theta e^{-\rho' t} + (1 - \theta) e^{-\rho'' t}, \qquad \theta e^{-\psi \rho' t} + (1 - \theta) e^{-\rho'' t}, \qquad (5.27)$$

where $0 < \theta < 1$ and if ρ' and ρ'' are appreciably different, plots of log survivor function versus t will expose what is happening.

An alternative rather more empirical approach for accelerated onset is to set out an initial perturbation of the hazard $h_0(t)$ with total integral zero, so that once this perturbation has been survived, the survivor functions are identical. A simple version is to propose hazard functions for the two groups of

$$h_0(t), \qquad h_0(t) + \alpha e^{-\gamma t}(t - 1/\gamma). \qquad (5.28)$$

For $\gamma t \gg 1$, hazard and survivor functions are identical. With a system more complex than the comparison of two groups, one or both of α and γ could be taken as functions of z.

Often it would be sensible to combine such an effect of accelerated onset with some modification of the whole hazard, e.g. by considering as the two hazards

$$h_0(t), \qquad \psi h_0(t) + \alpha e^{-\gamma t}(t - 1/\gamma). \qquad (5.29)$$

The special cases $\psi = 1, \alpha = 0; \psi = 1, \alpha \neq 0; \psi \neq 1, \alpha = 0$ all have direct interpretations.

5.7 Treatments with a transient effect

A situation rather similar to that of Section 5.6 arises when a treatment effect, or more generally influence of explanatory variables, is likely to be transient, i.e. applying only for small values of t. In the absence of a specific model based on a theoretical analysis of the system, such transient treatment effects can be represented empirically in various ways. Perhaps the simplest is to take, in the case of two groups, hazards

$$h_0(t), \qquad \exp(\beta_1 + \beta_2 e^{-\gamma t}) h_0(t). \qquad (5.30)$$

If $\beta_2 = 0$ we recover the simple proportional hazards model, whereas if $\beta_1 = 0, \gamma > 0$, the hazard for individuals in group 1 eventually reverts to that for group 0 and, in an appropriate context, individuals who survive until their hazard is very close to $h_0(t)$ can be regarded as 'ultimately cured'.

5.8 Discussion

In the previous sections we have outlined what may seem a bewildering variety both of broad types of model and of minor variants. Each is expressed in terms of a survivor function $\mathcal{F}_0(t)$ holding under some standard conditions $z = 0$ and of a modifying factor specified in various ways. The function $\mathcal{F}_0(t)$ may itself be specified parametrically; see Chapter 2.

If interest lies in the qualitative effect on failure time of various explanatory variables, choice of a model may not be critical. On the other hand, if interest lies in relatively subtle aspects of the dependence, or in discrimination between alternative specifications, a large amount of high-quality data is likely to be necessary. In some cases, especially in the physical sciences, there may be some special theory to guide the choice of model.

Bibliographic notes, 5

The parametric accelerated life model has several links with models widely used in other types of application, especially when $\psi(z; \beta)$ $= e^{\beta^T z}$. For it is then a log linear model, i.e. a linear regression model for log T, although in general with nonnormal error. For other forms of ψ, a nonlinear regression model results. The central assumption (5.4) for random variables is in the spirit of Fraser's (1968, 1979) structural inference.

Q–Q plots are described by Wilk and Gnanadesikan (1968) and used for nonparametric inference in the two-sample problem by Doksum and Sievers (1976).

In association with the exponential distribution, regression models of various forms were considered by Feigl and Zelen (1965), Cox and Snell (1968) and in a rather general log linear form by Glasser (1967). The proportional hazards model in a general form with time-dependent explanatory variables was given by Cox (1972) and the relation with the accelerated life model stated without proof; see Oakes (1981) for a review of subsequent developments. Non-multiplicative hazard-based models are studied by Aranda-Ordaz (1980).

While the qualitative ideas of accelerated onset and transient effects are well known, explicit models do not appear to have been formulated before.

Further results and exercises, 5

5.1. Find from (5.11) the time-dependent accelerated life version of the two-sample problem with $\mathscr{F}_1(t) = \mathscr{F}_0[(\rho t)^\kappa]$. Suggest how to find the function $\xi(t)$ 'nonparametrically' from the two survivor functions. Formulate three-sample, or more generally multi-sample, versions of the same problems.

5.2. Prove that a two-term representation (5.15) of two hazard functions $h_0(t)$ and $h_1(t)$ in terms of two underlying unobserved types of failure following accelerated life models is always possible if h_0 and h_1 are both of the Gompertz–Makeham form.

5.3. Examine the accelerated onset model (5.26) and (5.27) in which both component survivor functions are exponential. How could consistency with (5.26) be checked and parameters estimated graphically?

5.4. Suppose that, in the two-sample problem, the survivor functions satisfy the proportional odds model

$$\frac{\mathscr{F}_1(t)}{1 - \mathscr{F}_1(t)} = \psi \, \frac{\mathscr{F}_0(t)}{1 - \mathscr{F}_0(t)} \, .$$

(a) Show that the ratio of hazard functions $h_1(t)/h_0(t) \to 1$ as $t \to \infty$.

(b) Suppose that there is no censoring, and that survival is dichotomized at a point t_0, so that survivors beyond t_0 are counted as 'positives', failures before t_0 as 'negatives'. Show that the corresponding 2×2 contingency table has odds ratio ψ. How may this result be used (i) to estimate ψ and (ii) to assess graphically the goodness of fit?

(c) Show that if $\mathscr{F}_0(t)$ has the log logistic form then so does $\mathscr{F}_1(t)$.

(d) Investigate the general formulation where ψ is a function of explanatory variables, $\psi = \psi(z)$.

[Plackett, 1965; Clayton, 1974; McCullagh, 1980; Bennett, 1983]

5.5. Consider two groups with hazards $h_0(t), h_1(t)$ and survivor functions $\mathscr{F}_0(t), \mathscr{F}_1(t)$. Examine the connection between the crossings of the two hazard functions and the crossing of the two survivor functions, noting in particular the following:

(a) if the continuous survivor functions cross once, the hazards must cross at least once;

(b) it is possible that $\mathscr{F}_1(t) > \mathscr{F}_0(t)$ for all $t > 0$ and yet for the hazards to cross very often;

(c) if both groups consist of a mixture of 'short' survivors with possibly different distributions in the two groups and of 'long' survivors with the same distribution in the two groups, then the hazards, but not in general the survivor functions, are equal for large t.

5.6. Suppose that in comparing two groups the failure times in one group are almost all less than t' whereas those in the other group almost all exceed t'. What can be concluded about the two hazard functions; can the proportional hazards model be refuted? Is there any special difficulty in checking whether the accelerated life model holds?

5.7. For the accelerated life log logistic model of Section 5.2(vii) and Exercise 5.4, compare in the uncensored case the asymptotic efficiencies of full maximum likelihood estimation of the parameter β and that of maximum likelihood estimation of the logistic binary response model formed by considering only survival or failure at a fixed time t_0.

5.8. Suppose that $h_0(t)$ and $h_1(t)$ are continuous bounded hazard functions satisfying, for some ψ, $0 < \psi < 1$, and all t,

$$h_0(t) > h_1(t) > \psi h_0(\psi t).$$

Show that a representation of the form (5.15) with $l = 2$ in terms of nonnegative component hazard functions $h_{01}(t)$ and $h_{02}(t)$ is possible, and determine these functions in terms of $h_0(t)$ and $h_1(t)$.

5.9. By using an argument like that of Section 5.3(ii), show that the proportional odds model

$$\frac{\mathscr{F}_1(t)}{1 - \mathscr{F}_1(t)} = \psi \, \frac{\mathscr{F}_0(t)}{1 - \mathscr{F}_0(t)}$$

is an accelerated life model if and only if $\mathscr{F}_0(t)$ has the log logistic form.

Develop a similar characterization of the log normal distribution in terms of the probit model

$$\Phi^{-1}\{\mathscr{F}_1(t)\} = \psi + \Phi^{-1}\{\mathscr{F}_0(t)\},$$

where $\Phi(\,.\,)$ is the standard normal distribution function.

CHAPTER 6

Fully parametric analysis of dependency

6.1 Introduction

In Chapter 5 we reviewed some models representing the dependency of failure time on a vector z of explanatory variables. It is often useful to consider such models as formed from a baseline distribution holding under the standard conditions, $z = 0$, plus a specification of the modification induced by nonzero z.

In the present chapter, we suppose that both parts of the model are determined by a limited number of unknown parameters. We call such a specification fully parametric, to be contrasted with wholly or partially nonparametric formulations to be studied later. That is, we suppose that survivor function, density and hazard have the form

$$\mathscr{F}(t; z, \theta), \qquad f(t; z, \theta), \qquad h(t; z, \theta),$$

where θ is a parameter vector. Often θ^{T} is usefully partitioned as $(\phi^{\mathrm{T}}, \beta^{\mathrm{T}})$, where ϕ refers to the baseline survivor function and β to the explanatory variables. Usually, at each stage of application, there will be a small number of parameters of direct interest, in general discussion denoted by ω, the remaining nuisance parameters being denoted by λ.

No rigid rules can be laid down to dictate when to use a fully parametric approach. The following points are, however, relevant:

(i) in relatively complex problems, some parameterization is needed to provide a basis for concise summarization;

(ii) simple comparisons involving relatively large amounts of data are often best achieved by the informal comparison of a number of empirical survivor functions, involving, that is, a nonparametric view;

(iii) when censoring is present and relatively complex dependencies are to be studied, the proportional hazards model is the only one that in any generality allows an analysis without parametric formulation of the baseline survivor function;

(iv) when primary interest focuses on the explanatory variables, the precise specification of the baseline survivor function is often not critical;

(v) some formal or informal check on parametric formulation is always desirable, often examining separately the two component parts of the model;

(vi) the sensitivity of conclusions to observations extreme, either in failure time or in the space of z, needs examination; this is often best done by careful inspection.

For the analysis of fully parametric models, we concentrate on methods based on the likelihood function. Iterative numerical solution of the maximum likelihood equations is nearly always involved and the availability of suitably flexible computer programs is thus crucial. In Section 6.3(ii), however, we consider rather simpler methods for the accelerated life model without censoring.

For observations of n mutually independent individuals with explanatory variables z_i and failure or censoring time $x_i (i = 1,...,n)$, the log likelihood is, under the usual assumption about uninformative censoring,

$$l(\theta; x) = \sum_u \log f(x_i; z_i; \theta) + \sum_c \log \mathscr{F}(x_i; z_i; \theta)$$
$$= \sum_u \log h(x_i; z_i; \theta) + \sum \log \mathscr{F}(x_i; z_i; \theta), \qquad (6.1)$$

where $\sum_u$, $\sum_c$ and $\sum$ denote respectively summation over uncensored (i.e. failed) individuals, over censored individuals and over all individuals.

6.2 Exponential distribution

(i) Form of likelihood

We consider first models in which z is constant for each individual and in which failure time is exponentially distributed with rate parameter depending on the explanatory variable. This is a special case of an accelerated life model. We write $\rho(z_i; \beta)$ for the rate parameter of the ith individual, where β is unknown. Then, provided that the individuals are mutually independent, the log likelihood is

$$\sum_u \log \rho(z_i; \beta) - \sum \rho(z_i; \beta) x_i, \qquad (6.2)$$

where the sums are respectively over uncensored individuals and over all individuals.

The gradient vector, the matrix of second derivatives and the equations for the maximum likelihood estimate can thus be written down. If the function ρ is positive only for certain combinations of z and β, care may be needed in case the maximum is achieved on the boundary of the parameter space. This difficulty is avoided by the common choice

$$\rho(z;\beta) = e^{\beta^{\mathrm{T}}z};\tag{6.3}$$

in this it is usually convenient to suppose that the vector z includes a component z_0 equal to one for all individuals, so that e^{β_0} is the baseline rate for an individual with all the other components of z zero. Alternatively, we replace (6.3) by $\rho_0 e^{\beta^{\mathrm{T}}z}$.

With the choice (6.3) the log likelihood is

$$\sum_{\mathrm{u}}\beta^{\mathrm{T}}z_i - \sum e^{\beta^{\mathrm{T}}z_i}x_i.\tag{6.4}$$

On forming the second derivatives of (6.4) with respect to the components of β, we have that the Fisher information for the estimation of β has (r,s)th component

$$\sum z_{ir}z_{is}e^{\beta^{\mathrm{T}}z_i}E(X_i).$$

If there is no censoring, $E(X_i)e^{\beta^{\mathrm{T}}z_i} = 1$, so that the information matrix is $z^{\mathrm{T}}z$, where z is the $n \times q$ matrix of explanatory variables. If there is censoring, it is easily shown (see Exercise 6.1) that $E(X_i)e^{\beta^{\mathrm{T}}z_i} = 1 - \pi_i$, where π_i is the probability that the ith individual is censored. Thus, if $\pi = \mathrm{diag}(\pi_1,\ldots,\pi_n)$, the information matrix is

$$z^{\mathrm{T}}(I^* - \pi)z,\tag{6.5}$$

where I^* is the $n \times n$ identity matrix.

(ii)   Some 'exact' results

Although the topic is of largely theoretical interest, we now examine briefly 'exact' confidence intervals and tests. Except in very exceptional circumstances, these are not available for censored data and therefore we concentrate on the uncensored case. There are two different circumstances in which 'exact' analysis is in principle possible.

First, suppose that we have a linear model for the canonical parameter of the exponential distribution, i.e.

$$\rho(z;\beta) = \beta^\mathrm{T} z. \tag{6.6}$$

More precisely, the right-hand side of (6.6) is $(\beta^\mathrm{T} z)^+ = \max(\beta^\mathrm{T} z, 0)$.
The log likelihood when there is no censoring is

$$\sum \log(\beta^\mathrm{T} z_i) - \beta^\mathrm{T} \sum z_i t_i,$$

showing that we have a full exponential family with minimal sufficient statistic $\sum z_i t_i$. To test a hypothesis about a component of β, or a linear combination of components, we use the conditional distribution of the relevant component of $\sum z_i t_i$ given the remaining components. This distribution has usually to be approximated via an Edgeworth or saddle-point approximation.

Secondly, suppose that (6.3) holds, so that a linear regression model, namely

$$\log T_i = \mu_0 - \beta^\mathrm{T} z_i + \varepsilon_i \tag{6.7}$$

holds for $\log T$. Note that alternatively, as remarked below (6.3), we may include a constant component in the vector of explanatory variables to capture the effect of μ_0. In (6.7) the ε_i are, except for a constant, the logs of exponentially distributed random variables and hence have an extreme value distribution. Because of the invariance of (6.7) under an appropriate group of transformations, the residuals of $\log T_i$ from the least-squares regression have a fixed distribution independent of (μ, β) and hence are ancillary statistics. The conditional distribution of the least-squares estimates given the ancillary statistics is the frame for inference about (μ, β). Numerical procedures for converting this into a practically useful procedure do not seem to be available.

It is possible to analyse (6.7) by standard unconditional least-squares procedures. This involves fairly appreciable loss of efficiency; see Section 6.3(ii).

Note that very small values of T become extreme values of $\log T$ and may have an undue influence on the conclusions, especially of a least-squares analysis. Such influence is particularly undesirable if either small values of T are subject to relatively substantial recording errors, or if interest is focused primarily on the long failure times.

(iii) Two-sample problem

The simplest special case is the two-sample problem with interest concentrated on the ratio of the rate parameters. Write λ and $\omega\lambda$ for the rates in the two groups, so that the log likelihood (6.2) or (6.4) becomes

$$l(\omega, \lambda) = (d_0 + d_1) \log \lambda + d_1 \log \omega - \lambda t_0 - \lambda \omega t_1,$$

where d_0, d_1 are the numbers of failures and t_0, t_1 the observed total times at risk in the two groups. For given ω, the maximum likelihood estimate of λ is

$$\hat{\lambda}_\omega = (d_0 + d_1)/(t_0 + \omega t_1),$$

whereas the unconstrained maximum likelihood estimates are $\hat{\lambda} = d_0/t_0, \hat{\omega} = (d_1/t_1)(d_0/t_0)^{-1}$. Thus the maximum likelihood ratio statistic for testing $\omega = \omega_0$ is

$$W(\omega_0) = 2[l(\hat{\omega}, \hat{\lambda}) - l(\omega_0, \hat{\lambda}_{\omega_0})]$$

$$= 2\left[d_0 \log\left(\frac{d_0}{t_0}\right) + d_1 \log\left(\frac{d_1}{t_1}\right)\right.$$

$$\left. - (d_0 + d_1) \log\left(\frac{d_0 + d_1}{t_0 + \omega_0 t_1}\right) - d_1 \log \omega_0 \right]. \qquad (6.8)$$

A confidence region for ω follows from (3.6).

Now if d_0 and d_1 individuals are studied until all have failed, the corresponding total times at risk, T_0 and T_1, are independent and $2\lambda T_0$ and $2\lambda \omega T_1$ have chi-squared distributions with $2d_0$ and $2d_1$ degrees of freedom. Thus

$$F = \frac{\omega T_1/d_1}{T_0/d_0} \qquad (6.9)$$

has the standard variance-ratio distribution with $(2d_1, 2d_0)$ degrees of freedom. Significance tests and confidence tests for ω follow immediately. In fact (6.8) gives a close approximation to these limits, suggesting that (6.9) should be used to compare two rates when censoring is present and other 'stopping rules' are employed, taking d_0 and d_1, t_0 and t_1 to be the observed numbers of failures (fixed or random) and times at risk (fixed or random).

(iv)   Time-dependent explanatory variables

If the explanatory variables are time-dependent, the form of the log likelihood is more complicated, the survivor function being by (2.7)

$$\mathcal{F}(t; z) = \exp\left(- \int_0^t \rho[z(u); \beta]\, du \right)$$

and the log likelihood thus

$$\sum_u \log \rho[z_i(t_i);\beta] - \sum \int_0^{x_i} \rho[z_i(u);\beta]\,du. \qquad (6.10)$$

Except in simple cases such as when the z_i are step functions, the integrals in (6.10) have to be evaluated numerically; in any case, time-dependent explanatory variables will often be available only at discrete time points, so that interpolated values will be needed.

6.3 Accelerated life model

(i) Form of likelihood

As noted in Section 6.2, the exponentially based model with fixed explanatory variables is a special case of the accelerated life model. If that model is to be used with censored data, it is helpful if both density and survivor function can be computed in reasonably explicit form. For this reason, probably the three most important special distributions are the Weibull, the Gompertz–Makeham and the log logistic. In the kind of application considered in this book, it is the first and last of these that have been most used, the last partly because it allows for a maximum in the hazard function.

(ii) Uncensored case

When there are no censored observations, or when censoring is negligible, it is helpful to consider a regression model for log T. When the effect of the explanatory variables is represented by the factor $e^{\beta^T z}$, the regression equation has the log linear form

$$\log T_i = \mu - \beta^T z_i + \varepsilon_i \,;$$

under log logistic and Weibull (and of course other) distributions of failure time, the ε_i's are independent and identically distributed with unknown variance and known distributional form. The 'exact' argument of Section 6.2(ii) is modified because of the presence of the unknown standard deviation. The conditioning statistic becomes the set of standardized residuals, i.e. residuals divided by the root-mean-square residual.

For many purposes, an ordinary least-squares analysis will be adequate when there is no censoring. The asymptotic relative efficiency of least squares relative to maximum likelihood can be calculated. Provided that the model contains an unknown 'constant' term, this asymptotic relative efficiency depends neither on the dimension of β nor on the values of the explanatory variables. The value of efficiency is an average over the values of the ancillary statistic (standardized residuals), some configurations of which will be relatively favourable and others relatively unfavourable.

If least squares on logs is used when the distribution is exponential, the asymptotic relative efficiency is 0.61; for a gamma distribution with $\kappa = 5$, the corresponding value is 0.90. Further values are given by Cox and Hinkley (1968) and Pereira (1978), the latter giving also the asymptotic relative efficiency when maximum likelihood fitting of the wrong form of distribution is used.

(iii) More on the likelihood

We now discuss in a little more detail the log likelihood functions for two common cases, concentrating on the special representation in which the function $\psi(z; \beta)$ relating failure time T to failure time T_0 at $z = 0$, $T = T_0/\psi(z; \beta)$, has the form $\psi(z; \beta) = e^{\beta^T z}$.

For the Weibull distribution, the survivor function at $z = 0$ is

$$\mathscr{F}_0(t) = \exp[-(\rho t)^\kappa], \tag{6.11}$$

whereas for the log logistic

$$\mathscr{F}_0(t) = [1 + (\rho t)^\kappa]^{-1}. \tag{6.12}$$

It is again convenient to introduce into the vector z of explanatory variables a component z_0 identically one, writing $\rho = e^{\beta_0}$, so that the forms of $\mathscr{F}(t; z)$ corresponding to (6.11) and (6.12) are respectively

$$\exp(- e^{\kappa \beta^T z} t^\kappa), \qquad (1 + t^\kappa e^{\kappa \beta^T z})^{-1}.$$

Therefore, the log likelihood is for the Weibull distribution

$$-\sum e^{\kappa \beta^T z_i} x_i^\kappa + \sum_u [\kappa \beta^T z_i + (\kappa - 1)\log t_i + \log \kappa]$$

and for the log logistic

$$-\sum \log(1 + x_i^\kappa e^{\kappa \beta^T z_i})$$
$$+ \sum_u [(\kappa - 1)\log t_i + \log \kappa + \kappa \beta^T z_i - \log(1 + t_i^\kappa e^{\kappa \beta^T z_i})].$$

Maximization of the log likelihood and the calculation of associated quantities required for asymptotic procedures is now, in principle, straightforward.

6.4 Proportional hazards model

In the simplest form of the proportional hazards model, the survivor function, density and hazard are respectively

$$[\mathscr{F}_0(t)]^{\psi(z)}, \qquad \psi(z)[\mathscr{F}_0(t)]^{\psi(z)-1} f_0(t), \qquad \psi(z)h_0(t),$$

where $\mathscr{F}_0(t)$, etc., refer to the baseline conditions $z = 0$ and z is a vector of fixed explanatory variables. If $\psi(z)$ is represented parametrically as $\psi(z;\beta)$ and $\mathscr{F}_0(t)$ also is described parametrically as $\mathscr{F}_0(t;\phi)$, the likelihood for n independent individuals can be written down and it is again especially convenient if both $\mathscr{F}$ and f can be computed easily.

If the explanatory variables are time-dependent, the survivor function is

$$\exp\left(- \int_0^t \psi[z(u)]h_0(u;\phi)\, du \right)$$

and the difficulties of computation commented on in connection with the exponential case (6.10) are accentuated.

6.5 Sensitivity analysis

It is always important to examine the sensitivity of conclusions to the various assumptions implicit or explicit in the analysis. Often this is best done *ad hoc* for each set of data, although some broad comments can be made.

Misspecification of the baseline distribution may have relatively little effect on the estimation of the relative importance of the various explanatory variables, although the estimated standard error of regression coefficients will be too high or too low if the dispersion of the baseline distribution is less or more than that specified: of course, efficiency of estimation, too, is affected by distributional shape. Misspecification of the effect of explanatory variables is likely to lead to some systematic variation being absorbed into the baseline distribution and hence to the dispersion of that distribution being overestimated.

Nonindependence of the responses of different individuals can take various forms.

(a) If the assumed model is an adequate representation of the marginal density of failure time, methods based on independence will usually give sensible point estimates; if the association between individuals is predominantly positive, the true standard error of the parameter estimates may be much greater than that calculated assuming independence.

(b) If the dependence arises from grouping of individuals into distinct sets, it may be possible to absorb the lack of independence by assigning specific parameters to each set to represent between-set differences. This procedure is, however, dangerous if a large number of such parameters have to be introduced, unless they can be 'eliminated' from the likelihood by a conditioning or similar argument, for the usual good properties of maximum likelihood estimates can no longer be counted on when the number of parameters fitted is relatively large.

(c) If the dependence present is of intrinsic interest, it will usually be best to consider one of the special models of Chapter 10, at least if appreciable censoring is present.

(d) If the dependence is a temporal or spatial one, it may be possible to adapt some of the usual time-series models, but little work of this kind specifically connected with survival data seems to have been done.

6.6 Some less formal methods

The previous sections have concentrated on the fitting of specific parametric models. This leaves open two related important issues that arise in general in connection with highly specified models. These are procedures for developing appropriate models and for examining the adequacy of models that have been fitted or suggested.

One reasonably general procedure for preliminary analysis is to group the individuals into sets with approximately constant values of the more important explanatory variables. Nonparametric estimates of survivor or hazard function can then be obtained for each set and examined graphically or more formally for common features; for instance, if plots of log hazard versus log time produce parallel straight lines, a Weibull distribution of constant index is called for. Then as a second stage, the approximate form of dependence on the explanatory variables can be examined, leading, if formal model fitting is desirable, to such a model.

To proceed in a reverse direction, there are several possibilities.

Suppose that one of the models of this chapter has been fitted. We may examine its adequacy by

(a) fitting a more complex model containing additional parameters designed to test for specific types of departure in distributional form, or in the nature of the dependence on the explanatory variables;

(b) residuals can be defined for each individual that should, if the model is adequate, behave like a mildly constrained set of observations from some fixed and known distribution often, the unit exponential distribution is chosen for this purpose;

(c) particular attention may be focused on especially influential observations, i.e. on observations whose omission or inclusion has a strong effect on the main conclusions—these often correspond either to individuals with very long or very short failure times, or to individuals extreme in the space of the explanatory variables.

The general idea behind the definition of residuals with a unit exponential distribution, is that if T_i has survivor function $\mathscr{F}(t; z_i; \theta)$, then $\mathscr{F}(T_i; z_i; \theta)$ is uniformly distributed and $-\log \mathscr{F}(T_i; z_i; \theta)$ has a unit exponential distribution. If the full parameter vector is estimated by maximum likelihood, we thus define the residual to be

$$R_i = -\log \mathscr{F}(T_i; z_i; \hat{\theta}).$$

If the ith individual is censored, so too is the corresponding residual and thus in general we obtain a set of uncensored and a set of censored residuals.

For examining distributional form, the residuals can be analysed. For examining the dependency on explanatory variables, whether new or already in the model, direct plotting can be used with distinguished marking of the censored values.

Bibliographic notes, 6

Most of the key references have been given in the previous chapter. Cox and Lewis (1966) have described methods associated with the exponential distribution in the slightly different context of the analysis of series of events (point processes). Reid (1981a) obtained the influence function for some simple methods of analysis, and in as yet unpublished work has given the corresponding functions for complex analyses; these are relevent for diagnostic tests for outliers, for example. Zelen and Dannemiller (1961) and Vaeth (1979) have discussed the sensitivity of analyses to the assumption of exponential form. Crowley and Hu (1977) calculated residuals for a proportional hazards model developing a general definition of Cox and Snell (1968); see also Kay (1977) and Lagakos (1981).

Further results and exercises, 6

6.1. Prove that if T is exponentially distributed with parameter ρ and censoring is at a fixed time c, then the probability of censoring is $\pi = e^{-\rho c}$ and if $X = \min(T, c)$,

$$\rho E(X) = 1 - \pi.$$

Hence show that the same result holds when the censoring time is a random variable with an arbitrary distribution.

6.2. Consider two independent samples subject to censoring, the corresponding survivor functions being $\mathscr{F}_1(t)$ and $\mathscr{F}_2(t)$. Extend the argument of Section 4.3 to obtain a likelihood ratio test of the null hypothesis $\mathscr{F}_1(t_0) = \mathscr{F}_2(t_0)$, where t_0 is a preassigned time, the survivor functions being otherwise arbitrary. Examine the asymptotic relative efficiency of the procedure when in fact the distributions are exponential.

CHAPTER 7

Proportional hazards model

7.1 Introduction

This chapter considers the simple form of the proportional hazards model introduced in Section 5.3, namely

$$h(t,z) = \psi(z;\beta)h_0(t), \tag{7.1}$$

in which the explanatory vector z does not change over time for any individual. We shall assume for the moment that the survival times have continuous distributions and are recorded exactly, so that there is no possibility of ties. Three parameterizations of ψ may be considered, namely the log linear form $\psi(z;\beta) = e^{\beta^T z}$, which for good reasons has become the most popular, the linear form $\psi(z;\beta) = 1 + \beta^T z$, and the logistic, $\psi(z;\beta) = \log(1 + e^{\beta^T z})$. Discrimination among these forms may be achieved by fitting an augmented family. For example, the family

$$\psi(z,\beta;\kappa) = (1 + \kappa\beta^T z)^{1/\kappa}$$

includes the linear and log linear models as special cases, $\kappa = 1$ and $\kappa \to 0$ respectively.

7.2 The likelihood function

We now consider inferences about β when the hazard function $h_0(t)$ is completely unknown. It is convenient to treat first the case of no censoring. Let $\tau_1 < \tau_2 < \ldots < \tau_n$ denote the ordered failure times of the n individuals and let $\mathscr{I}_j$ be the label of the subject who fails at τ_j. Thus $\mathscr{I}_j = i$ if and only if $t_i = \tau_j$. Write $\mathscr{R}(\tau_j) = \{i : t_i \geqslant \tau_j\}$ for the risk set just before the jth ordered failure and r_j for its size. These definitions are illustrated in Fig. 7.1.

The basic principle of the derivation of the likelihood is as follows. The $\{\tau_j\}$ and $\{\mathscr{I}_j\}$ are jointly equivalent to the original data, namely the unordered failure times t_i. In the absence of knowledge of $h_0(t)$, the

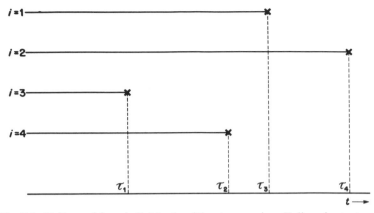

Fig. 7.1. Failure of four individuals without censoring. Failure instants $\tau_1,\dots,\tau_4$. Risk sets are $\mathscr{R}(\tau_1) = \{1, 2, 3, 4\}$; $\mathscr{R}(\tau_2) = \{1, 2, 4\}$; $\mathscr{R}(\tau_3) = \{1, 2\}$; $\mathscr{R}(\tau_4) = \{2\}$. Contribution to relation given below (7.3)

τ_j can provide little or no information about β, for their distribution will depend heavily on $h_0(t)$. As an extreme example, $h_0(t)$ might be identically zero except in small neighbourhoods of the τ_j. Attention must, therefore, be focused on the $\mathscr{I}_j$. In the present case, their joint distribution $p(i_1, i_2, \dots, i_n)$ over the set of all possible permutations of $(1, 2, \dots, n)$ can be derived explicitly.

The conditional probability that $\mathscr{I}_j = i$ given the entire history

$$\mathscr{H}_j = \{\tau_1, \tau_2, \dots, \tau_j, i_1, i_2, \dots, i_{j-1}\}$$

up to the jth ordered failure time τ_j is easily written down. It is the conditional probability that i fails at τ_j given that one individual from the risk set $\mathscr{R}(\tau_j)$ fails at τ_j, which is simply

$$\frac{h_i(\tau_j)}{\sum_{k \in \mathscr{R}(\tau_j)} h_k(\tau_j)} = \frac{\psi(i)}{\sum_{k \in \mathscr{R}(\tau_j)} \psi(k)}, \tag{7.2}$$

the baseline hazard function $h_0(\tau_j)$ cancelling because of the multiplicative assumption (7.1). For notational convenience, $\psi(k)$ here denotes $\psi(z_k, \beta)$, that is, the multiplier ψ for the kth subject.

Although (7.2) was derived as the conditional probability that $\mathscr{I}_j = i$ given the entire history $\mathscr{H}_j$, in fact it is functionally independent of the $\tau_1, \tau_2, \dots, \tau_j$. It therefore equals $p_j(i | i_1, i_2, \dots, i_{j-1})$, the conditional distribution of $\mathscr{I}_j$ given only $\mathscr{I}_1 = i_1, \mathscr{I}_2 = i_2, \dots,$

$\mathcal{I}_{j-1} = i_{j-1}$. The joint distribution $p(i_1, i_2, \ldots, i_n)$ can therefore be obtained by the usual chain rule for conditional probabilities as

$$p(i_1, i_2, \ldots, i_n) = \prod_{j=1}^{n} p_j(i_j | i_1, i_2, \ldots, i_{j-1})$$

$$= \prod_{j=1}^{n} \frac{\psi(i_j)}{\sum_{k \in \mathcal{R}(\tau_j)} \psi(k)}. \tag{7.3}$$

As an example, consider again the configuration of Fig. 7.1. Here, $\mathcal{I}_1 = 3, \mathcal{I}_2 = 4, \mathcal{I}_3 = 1$ and $\mathcal{I}_4 = 2$, so that the required probability is

$$p(3, 4, 1, 2) = \frac{\psi(3)}{\psi(1) + \psi(2) + \psi(3) + \psi(4)} \times \frac{\psi(4)}{\psi(1) + \psi(2) + \psi(4)}$$

$$\times \frac{\psi(1)}{\psi(1) + \psi(2)} \times \frac{\psi(2)}{\psi(2)}.$$

In the presence of censoring, a similar argument applies if it can be assumed that censorings can only occur immediately after failures. This requirement does conflict slightly with the model in which the censoring times are fixed constants, but can usually be viewed as a reasonable approximation, as the information about β contributed by an exact observed censoring time c_i will generally be small. The fixed censoring model can be handled explicitly through partial likelihood; see Chapter 8.

Suppose then, that there are d observed failures from the sample of size n, and let the ordered observed failure times be $\tau_1 < \tau_2 < \ldots < \tau_d$. As before, let $\mathcal{I}_j = i$ if subject i fails at τ_j, and let $\mathcal{R}(\tau_j) = \{i : t_i \geq \tau_j\}$ be the corresponding risk set with size r_j. Equation (7.2) follows exactly as before, where $\mathcal{H}_j$ now includes the censorings in $(0, \tau_j)$ as well as the failures, and the fact that no censoring can occur in (τ_{j-1}, τ_j) ensures that the risk set $\mathcal{R}(\tau_j)$, and hence the expression (7.2), does not depend on τ_j.

Combination of these conditional probabilities gives the overall likelihood

$$\text{lik} = \prod_{j=1}^{d} \frac{\psi(i_j)}{\sum_{k \in \mathcal{R}(\tau_j)} \psi(k)}. \tag{7.4}$$

We have termed this a likelihood rather than a probability, because terms that determine which individuals should be censored from among the survivors of each risk set have been omitted. As long as the

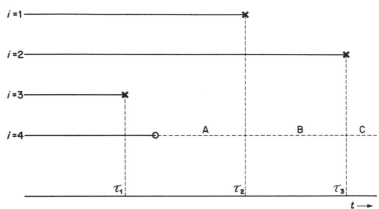

Fig. 7.2. Failure of four individuals with censoring. ×, failure; ○, censoring. Failure instants τ_1, τ_2, τ_3. Risk sets are $\mathscr{R}_1 = \{1, 2, 3, 4\}$; $\mathscr{R}_2 = \{1, 2\}$; $\mathscr{R}_3 = \{2\}$. Contribution to likelihood, (7.5). A, B, C possible position for failure time of censored individual.

censoring mechanism itself does not depend on β, these terms will not depend functionally on β and can be ignored for the purpose of likelihood inference about β. Censoring mechanisms that do depend on the same parameters as the failure mechanism will be discussed in Chapter 9.

Alternatively, (7.4) can be derived as the sum of all probabilities (7.3) consistent with the observed pattern of failures and censorings. The principle of the argument is illustrated in Fig. 7.2. The likelihood is

$$\text{lik} = \frac{\psi(3)}{\psi(1) + \psi(2) + \psi(3) + \psi(4)} \cdot \frac{\psi(1)}{\psi(1) + \psi(2)} \cdot \frac{\psi(2)}{\psi(2)}, \tag{7.5}$$

which is the same as the sum of the three terms corresponding to the possible positions A, B, C for τ_4 relative to τ_1, τ_2 and τ_3, namely

A: $\dfrac{\psi(3)}{\psi(1) + \psi(2) + \psi(3) + \psi(4)} \cdot \dfrac{\psi(4)}{\psi(1) + \psi(2) + \psi(4)} \cdot \dfrac{\psi(1)}{\psi(1) + \psi(2)} \cdot \dfrac{\psi(2)}{\psi(2)}$;

B: $\dfrac{\psi(3)}{\psi(1) + \psi(2) + \psi(3) + \psi(4)} \cdot \dfrac{\psi(1)}{\psi(1) + \psi(2) + \psi(4)} \cdot \dfrac{\psi(4)}{\psi(2) + \psi(4)} \cdot \dfrac{\psi(2)}{\psi(2)}$;

C: $\dfrac{\psi(3)}{\psi(1) + \psi(2) + \psi(3) + \psi(4)} \cdot \dfrac{\psi(1)}{\psi(1) + \psi(2) + \psi(4)} \cdot \dfrac{\psi(2)}{\psi(2) + \psi(4)} \cdot \dfrac{\psi(4)}{\psi(4)}$;

See Exercise 7.1 for the general argument.

7.3 The derivatives of the log likelihood

The proportional hazards form (7.1) is essential to the derivation of (7.4) but the precise functional form of $\psi(z;\beta)$ is not. In this section, we shall assume that for all $i, \psi(i)$ has first and second derivatives

$$\frac{\partial}{\partial \beta_r}\psi(i) = \psi_r(i), \qquad \frac{\partial^2}{\partial \beta_r \partial \beta_s}\psi(i) = \psi_{rs}(i).$$

Note that we do not actually need to assume that ψ can be expressed explicitly as a function of a covariate vector, although it almost always will be. It would suffice for the multiplier $\psi(i)$ for each subject i to be expressed as a function of the parameters β.

By a slight abuse of notation, (7.4) may be expressed in terms of the unordered observed failure times t_i, with corresponding risk sets $\mathcal{R}(t_i) = \mathcal{R}_i$, as

$$\mathrm{lik} = \prod_{i \in \mathcal{D}} \frac{\psi(i)}{\sum_{k \in \mathcal{R}_i} \psi(k)},$$

where $\mathcal{D}$ denotes the set of individuals failing. The corresponding log likelihood is

$$l = \sum_{i \in \mathcal{D}} \left[\log[\psi(i)] - \log\left(\sum_{k \in \mathcal{R}_i} \psi(k) \right) \right] = \sum_{i \in \mathcal{D}} l_i,$$

say. The derivatives of the contribution to l from the ith failure and its risk set are

$$\frac{\partial l_i}{\partial \beta_r} = \frac{\psi_r(i)}{\psi(i)} - \frac{\sum_{k \in \mathcal{R}_i} \psi_r(k)}{\sum_{k \in \mathcal{R}_i} \psi(k)}, \qquad (7.6)$$

$$\frac{\partial^2 l_i}{\partial \beta_r \partial \beta_s} = \frac{\psi_{rs}(i)}{\psi(i)} - \frac{\psi_r(i)\psi_s(i)}{[\psi(i)]^2} - \frac{\sum_{k \in \mathcal{R}_i} \psi_{rs}(k)}{\sum_{k \in \mathcal{R}_i} \psi(k)}$$

$$+ \frac{\sum_{k \in \mathcal{R}_i} \psi_r(k) \sum_{k \in \mathcal{R}_i} \psi_s(k)}{\left(\sum_{k \in \mathcal{R}_i} \psi(k) \right)^2}. \qquad (7.7)$$

The expectations of these quantities when i is sampled with probability proportional to $\psi(i)$ from the risk set $\mathscr{R}_i$ are of some interest. It is easily seen that

$$E(\partial l_i/\partial \beta_r) = 0,$$

as indeed, the derivative being a score function, must be so. On taking expectations of the second derivatives, the terms in ψ_{rs} cancel and

$$
\begin{aligned}
- E\left(\frac{\partial^2 l_i}{\partial \beta_r \partial \beta_s}\right) &= \frac{\sum \psi_r(k)\psi_s(k)/\psi(k)}{\sum \psi(k)} - \frac{\sum \psi_r(k)\psi_s(k)}{(\sum \psi(k))^2} \\
&= \mathrm{cov}\left(\frac{\psi_r(i)}{\psi(i)}, \frac{\psi_s(i)}{\psi(i)}\right) \\
&= \mathrm{cov}\left(\frac{\partial l_i}{\partial \beta_r}, \frac{\partial l_i}{\partial \beta_s}\right),
\end{aligned}
\tag{7.8}
$$

under this particular weighted sampling.

These expectations and covariances are taken over a single risk set, and are of course conditional on the composition of that risk set. Calculation of fully unconditional expectations would require a fuller specification of the censoring mechanism. The expectations just calculated can, however, be taken as conditional on the entire history of failures and censorings up to t_i, and this, in turn, allows a direct verification that the terms l_i of l do have some of the desirable properties of the increments of a log likelihood function. Further discussion of this point is postponed to Chapter 8.

It is easily seen that the matrix $-E(\partial^2 l_i/\partial \beta_r \partial \beta_s)$ must be non-negative definite, but that the matrix formed from the observed second derivatives, $-\partial^2 l_i/\partial \beta_r \partial \beta_s$, need not be, in general.

7.4 Log linear hazards

Important simplifications occur when $\psi(z, \beta)$ takes the log linear form $e^{\beta^T z}$. For then $\psi_r(i) = z_{ir}\psi(i)$ and $\psi_{rs}(i) = z_{ir}z_{is}\psi(i)$, where as usual, z_{ir} denotes the value of the rth component of the explanatory variable z on the ith subject. We have that

$$
\frac{\partial l_i}{\partial \beta_r} = z_{ir} - \frac{\sum_{k \in \mathscr{R}_i} z_{kr} e^{\beta^T z_k}}{\sum_{k \in \mathscr{R}_i} e^{\beta^T z_k}} = z_{ir} - A_{ir}(\beta),
\tag{7.9}
$$

say, the difference between the value of the explanatory variable on the failed subject, and the weighted average of the same variable over the corresponding risk set. Also

$$\frac{\partial^2 l_i}{\partial \beta_r \partial \beta_s} = -\frac{\sum\limits_{k \in \mathscr{R}_i} z_{kr} z_{ks} e^{\beta^T z_k}}{\sum\limits_{k \in \mathscr{R}_i} e^{\beta^T z_k}} + A_{ir}(\beta) A_{is}(\beta)$$

$$= -C_{irs}(\beta), \tag{7.10}$$

say.

Here, the observed and expected values of $\partial^2 l_i / \partial \beta_r \partial \beta_s$, taken over a single risk set are identical. Minus the matrix of second derivatives has the form of a covariance matrix under the weighted sampling scheme, and is nonnegative definite. Summing over all risk sets, we have that the score function $U(\beta)$ has rth component

$$U_r(\beta) = \sum_{i \in \mathscr{D}} \frac{\partial l_i}{\partial \beta_r} = \sum_{i \in \mathscr{D}} z_{ir} - A_{ir}(\beta), \tag{7.11}$$

and the information matrix $I(\beta)$ of negative second derivatives has elements

$$I_{rs}(\beta) = \sum_{i \in \mathscr{D}} C_{irs}(\beta). \tag{7.12}$$

Using the arguments outlined in the previous section, it may be shown that the $U_r(\beta)$ have zero expectation and covariance matrix $E[I(\beta)]$. Calculation of this expectation would require a fuller specification of the censoring mechanism than is normally available or desirable; in the analysis of a specific set of data it seems irrelevant to involve the times at which individuals, who in fact failed, would have been censored. In any case, recent work on asymptotic theory points to the use of observed information matrices and to tests and confidence interval procedures that use the likelihood as directly as possible. Thus, use of $I(\beta)$ rather than $E[I(\beta)]$ is appropriate, except in a theoretical analysis of competing procedures; see Section 3.3 for references.

The three main methods for statistical analysis based on log likelihoods, namely the likelihood ratio test, the score test and the direct use of the maximum likelihood estimates, have been reviewed in Section 3.3 and all are applicable here.

In the absence of censoring, or if the censoring mechanism is independent of the explanatory variables, an exact test of the null hypothesis $\beta = 0$ can be obtained by referring the score statistic

$$U(0) = \sum_{i \in \mathcal{D}} [z_i - A_i(0)] \tag{7.13}$$

to its permutation distribution. This is the distribution of $U(0)$ generated when the ordered failure times $\tau_1, \tau_2, \ldots, \tau_d$ and the sizes of the corresponding risk sets $r_1, r_2, \ldots, r_d$ are taken as fixed and the n values $z_1, z_2, \ldots, z_n$ of the explanatory variables are permuted randomly among the n subjects. For this is precisely the conditional distribution of $U(0)$ given $r_1, r_2, \ldots, r_d$ under the full model.

Some insight into the nature of the permutation distribution is given by a re-expression of $U(0)$ as

$$U(0) = \sum_{i=1}^{n} q_i z_i, \tag{7.14}$$

where

$$q_i = \delta_i - \sum_{j : \tau_j \leqslant t_i} 1/r_j$$

and $\delta_i = 0$ or 1 according as the ith individual is censored or uncensored. To demonstrate the equivalence of (7.13) and (7.14) note that

$$\sum_{i=1}^{n} z_i \left(\sum_{j : \tau_j \leqslant t_i} \frac{1}{r_j} \right) \equiv \sum_{j=1}^{d} \frac{1}{r_j} \sum_{i \in \mathcal{R}(\tau_j)} z_i.$$

By setting $z_i \equiv 1$ in this identity, we see that $\sum q_i = 0$, so that $E[U(0)] = 0$ under the permutation distribution. Also, the covariance matrix of $U(0)$ is

$$\tilde{I}(0) = \frac{1}{n-1} \left(\sum_{i=1}^{n} (z_i - \bar{z})(z_i - \bar{z})^{\mathrm{T}} \right) \left(\sum_{i=1}^{n} q_i^2 \right), \tag{7.15}$$

where $\bar{z} = \sum z_i / n$. This covariance matrix differs from $I(0)$ and is valid, as is the entire permutation theory, only under the assumption that the censoring mechanism is the same in the two groups.

In the absence of censoring, the ordered $q_i - \delta_i$ reduce to the so-called exponential ordered scores, e_{ni}, of Section 4.5.

7.5 An example

As a simple example of these techniques, we consider the data of Feigl and Zelen (1965) given in Table 1.2 on survival times of leukaemia patients. Using different methods, Feigl and Zelen reached two conclusions:

(i) the dependence of survival time on the concomitant variable (white blood count) is strong for the AG-positive group but quite weak for the AG-negative group;

(ii) for all values of the concomitant variable, the survival curve for the AG-positive group lies above that for the AG-negative group.

We fit models of the form

$$h_i(t) = \exp(\beta_1 z_1 + \beta_2 z_2 + \beta_3 z_3) h_0(t),$$

where $z_1 = 0$ (AG negative), $z_1 = 1$ (AG positive), $z_2 = \log(\text{WBC}) - 9.531$ and $z_3 = (z_1 - 0.5152) z_2$. The centring of the components z_2 and z_3 is partly for computational reasons and partly to aid in the interpretation. Four different models were fitted in addition to the 'null' model of no covariate effects, namely

(I) $\beta_1 \neq 0$, $\beta_2 = \beta_3 = 0$, corresponding to a difference in the failure times in the two groups, but no effect of the white blood count (in either group);

(II) $\beta_1 = 0$, $\beta_2 \neq 0$, $\beta_3 = 0$, corresponding to a dependence of failure time on white blood count, but no difference between groups;

(III) $\beta_1 \neq 0$, $\beta_2 \neq 0$, $\beta_3 = 0$, a model which allows both a difference between the groups and an effect of white blood count, but assumes that the two effects are additive on the logarithmic scale, so that the two factors act multiplicatively on the hazard function;

(IV) $\beta_1 \neq 0$, $\beta_2 \neq 0$, $\beta_3 \neq 0$, the most general model with interaction, allowing different levels in the two groups for dependence of failure time on white blood count.

The alert reader will notice several tied survival times in the data, for example four subjects have reported survival times of four weeks. Strictly, the continuous-time model (7.1) and the likelihood (7.4) are inappropriate. In the present analysis, an approximation first suggested by R. Peto is used to overcome the difficulty. If several failures are reported at the same time τ, each contributes a separate term of the form (7.2), with the same denominator, including all subjects in the risk set $\mathcal{R}(\tau)$. This approximation is used in many computer programs and is usually quite adequate. More sophisticated approximations are discussed in Section 7.6.

Table 7.1 shows the values of the log likelihood for the five models,

Table 7.1 *Analysis of Feigl and Zelen's (1965) data (Table 1.2)*

Terms	Log likelihood	Estimate	Standard error
—	−86.00	—	—
z_1	−82.23	−1.12	0.41
z_2	−81.66	0.40	0.14
z_1	−78.68	−1.02	0.42
z_2		0.36	0.14
z_1	−77.03	−1.14	0.43
z_2		0.40	0.14
z_3		0.50	0.28

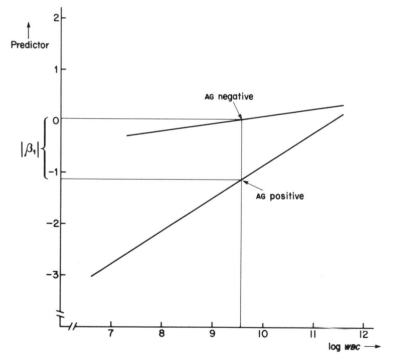

Fig. 7.3. Data of Feigl and Zelen. Log linear predictors fitted separately to two groups.

together with the estimated values of the parameters and their standard errors. The main effects of the group difference, z_1, and log(WBC) are both highly significant, when fitted either together or separately. The interaction term z_3 yields a change in log likelihood which fails to reach the 5% point of the chi-squared distribution:

$$2(78.68 - 77.03) = 3.30 < 3.84.$$

However, the numerical values of the coefficients expressing the dependence of $\beta_1 z_1 + \beta_2 z_2 + \beta_3 z_3$ on log (WBC) are

$$\hat{\beta}_2 - 0.515\hat{\beta}_3 = 0.14 \qquad \text{(AG negative)},$$
$$\hat{\beta}_2 + 0.485\hat{\beta}_3 = 0.64 \qquad \text{(AG positive)}.$$

The linear predictor $\beta_1 z_1 + \beta_2 z_2 + \beta_3 z_3$ is plotted against log (WBC) for the two groups in Fig. 7.3. Note that β_1 in this model represents the difference in the values of the linear predictor evaluated at $z_2 = 0$, i.e. for log (WBC) = 9.531. This plot supports the conclusions of Feigl and Zelen.

7.6 Discrete failure times: ties

As mentioned in the previous section, the likelihood (7.4) is not strictly appropriate for discrete survival times, which may involve ties. Two different approaches are possible here. If the timescale is genuinely discrete, the model (7.1) can be replaced by a discrete logistic model

$$\frac{h(t, z)}{1 - h(t, z)} = \psi(z; \beta)\frac{h_0(t)}{1 - h_0(t)},$$

where $h(t, z)$ now represents the discrete hazard

$$\text{pr}(T \leqslant t + 1 \mid T > t)$$

for an individual with explanatory variable z. To derive a likelihood function for β, let $\tau_1 < \tau_2 < \cdots < \tau_g$ denote the g distinct ordered failure times as before, and, additionally, let d_k denote the multiplicity of failures at τ_k. The history $\mathcal{H}_j$ now includes the multiplicities at all failure times up to and including τ_j. The conditional probability that items $i_1, i_2, \ldots, i_d$ fail from the risk set $\mathcal{R}(\tau_j)$ given $\mathcal{H}_j$ is then

$$\frac{\psi(i_1)\psi(i_2)\ldots\psi(i_d)}{\sum_{k \in s(j;d)} \psi(k_1)\psi(k_2)\ldots\psi(k_d)}, \tag{7.16}$$

where $s(j;d)$ denotes the set of all selections of $d = d_j$ items from the risk set $\mathcal{R}(\tau_j)$ of size $r_j = r$. Note that (7.16) represents the contribution from a single failure time. Because of the dependence on the multiplicity d_j, the product of all such terms is no longer a marginal likelihood of ranks, and the method of partial likelihood (Chapter 8) must be used to justify the asymptotic theory. Computation of the log likelihood and its derivatives is also very lengthy.

Some simplification is possible under the log linear form $\psi(z;\beta) = e^{\beta^T z}$. For then the full log likelihood is

$$ l = \sum_{j=1}^{g} \left[\beta^T s_j - \log\left(\sum_{k \in s(j;d)} e^{\beta^T s_{jk}} \right) \right], \tag{7.17} $$

where $s_j = z_{i_1} + z_{i_2} + \ldots + z_{i_d}$ is the sum of the vectors z_i over the individuals who actually fail at τ_j, each s_{jk} is the corresponding sum over a d tuple $(k_1, k_2, \ldots, k_d)$ of subjects who might have failed at τ_j. Fortunately, recursive algorithms make evaluation of the sums over $s(j;d)$ feasible even when r and d are too large to allow direct calculation of all terms.

The second approach to handling ties is to view them as arising out of the grouping of survival times that were actually generated from the continuous-time model. Unfortunately, this grouping does not yield the logistic model and the resulting likelihoods are different. In fact, no exact conditioning argument is available for the grouped model. Instead, we can sum all terms of the marginal likelihood (7.4) of ranks from the continuous model that are consistent with the observed data. If items $i = 1$, 2 are observed to fail at τ, from items $i = 1, 2, 3, 4$ at risk, the contribution from τ to this likelihood would be

$$ \frac{\psi(1)}{\psi(1)+\psi(2)+\psi(3)+\psi(4)} \frac{\psi(2)}{\psi(2)+\psi(3)+\psi(4)} $$
$$ + \frac{\psi(2)}{\psi(1)+\psi(2)+\psi(3)+\psi(4)} \frac{\psi(1)}{\psi(1)+\psi(3)+\psi(4)}. $$

Unfortunately, the sum does not simplify, and the corresponding log likelihood, being a sum of logarithms of terms of this form, is difficult to compute if several risk sets have values of r and d that are at all large.

The approximate likelihood mentioned and used in the previous section is obtained by augmenting all the sums in the denominators to include all items in the corresponding risk set. In the above example, this yields

$$\frac{2\psi(1)\psi(2)}{[\psi(1) + \psi(2) + \psi(3) + \psi(4)]^2}$$

and, more generally,

$$\frac{d!\psi(i_1)\psi(i_2)\ldots\psi(i_d)}{\left(\sum_{k \in \mathcal{R}(\tau)} \psi(k)\right)^d}, \tag{7.18}$$

which gives tractable log likelihoods. This approximation is widely used and is quite satisfactory except when the data exhibit heavy ties. The multiple counting of failed individuals does result in a conservative bias, however.

Two other simple estimators have been suggested to overcome this problem. These involve replacing the denominator of (7.18) by respectively

$$\prod_{j=1}^{d} \left(\sum_{k \in \mathcal{R}(\tau)} \psi(k) - \frac{(j-1)}{d} \sum_{k \in \mathcal{D}_j} \psi(k) \right), \tag{7.19}$$

$$\left[\sum_{k \in \mathcal{R}(\tau)} \psi(k) - \left(\frac{d-1}{2d}\right) \sum_{k \in \mathcal{D}_j} \psi(k) \right]^d, \tag{7.20}$$

where $\mathcal{D}_j$ is the set of individuals who fail at τ_j. None of these approximations as written allows for ties between reported censoring and failure times. The usual convention is to assume that all failures reported at any time τ precede any censorings reported at τ, so that the censored subjects contribute fully to the corresponding risk set(s). *Ad hoc* modifications of these approximate likelihoods to handle data where censored and uncensored failure times are grouped together in the same intervals can easily be derived.

In the absence of ties, i.e. if $d_j = 1$ for all j, all the preceding suggestions (7.17)–(7.20) give the same likelihood (7.4) as the continuous model.

With very heavy ties, so that the survival times are grouped into a small number of intervals, it becomes feasible to devote a separate nuisance parameter π_j to the (conditional) baseline survivor function for each interval and to carry out a full maximization of the log likelihood in both the π_j and the regression parameters β. For sensible results to be obtained, the total number of reported failures d must be much larger than the number g of grouping intervals. Asymptotic results require that the sample size $n \to \infty$ under fixed grouping intervals. For details, see Exercise 7.5.

7.7 The two-sample problem

We now apply the discrete logistic regression model to the comparison of two groups of survival times subject to censoring. Here the explanatory variable z has just one component which takes values 1 and 0 for groups 1 and 0 respectively. The log likelihood (7.17) is

$$l(\beta) = \beta \sum_{j=1}^{g} d_{1j} - \sum_{j=1}^{g} \log \left(\sum_{k \in s(j,d)} e^{\beta d_{1jk}} \right).$$

Here d_{1j} is the number of failures at τ_j in group 1 and d_{1jk} is the number of members of group 1 in the selection $k = (k_1, k_2, \ldots, k_d)$ of $d = d_j$ items from the risk set $\mathscr{R}(\tau_j)$.

The derivatives of the log likelihood are

$$U(\beta) = l'(\beta) = \sum_{j=1}^{g} [d_{1j} - A_j(\beta)],$$

and

$$-I(\beta) = l''(\beta) = -\sum_{j=1}^{g} V_j(\beta),$$

where $A_j(\beta)$ and $V_j(\beta)$ denote respectively the mean and the variance of d_{1jk} under weighted sampling without replacement from the risk set $\mathscr{R}(\tau_j)$, the probability of selection of $i \in \mathscr{R}(\tau_j)$ being proportional to $e^{\beta^T z_i}$.

For the score test of the null hypothesis $\beta = 0$ of no difference between the two groups, the values of $U(0)$ and $I(0)$ are needed. The sampling scheme defining $A_j(\beta)$ and $V_j(\beta)$ reduces to simple random sampling without replacement if $\beta = 0$, so that $A_j(0)$ and $V_j(0)$ are just the mean and variance of a hypergeometric random variable, namely

$$A_j(0) = d_j r_{1j}/r_j,$$

$$V_j(0) = d_j \frac{r_{0j} r_{1j} (r_j - d_j)}{r_j^2 (r_j - 1)}, \tag{7.21}$$

where r_{1j} and r_{0j} are the sizes of the risk sets in the two groups. The score test, which compares $U(0)$ to its estimated variance $I(0)$ is often called the log rank test, because of the relation to the exponential ordered scores indicated in Section 7.4.

The test can also be obtained formally by setting up a separate 2×2 contingency table at each failure time, with rows corresponding to group membership and columns to failure or survival, and carrying out the combined test for association according to the method of Mantel and Haenszel.

If Peto's approximate likelihood is used, the factor $(r_j - d_j)/(r_j - 1)$ no longer appears in $V_j(0)$ and the resulting variance estimate $I(0)$ is conservative. An even simpler test is sometimes used, based on the chi-squared statistic, on one degree of freedom, of

$$\chi^2 = \frac{(D_1 - E_1)^2}{E_1} + \frac{(D_0 - E_0)^2}{E_0},$$

(or a continuity corrected version), where $D_1 = \sum d_{1j}$ and $D_0 = \sum d_{0j}$ are the total numbers of failures in groups 1 and 0 and E_1 and E_0 are the corresponding expectations. Thus $F_1 = \sum_j d_j r_{1j}/r_j$. Since $(D_1 - E_1) - (D_0 - E_0) = U(0)$, this test is actually equivalent to the use of the variance approximation

$$\left(\frac{1}{E_0} + \frac{1}{E_1} \right)^{-1} = \frac{E_0 E_1}{E_0 + E_1}$$

in place of $I(0)$. This introduces further conservatism.

Table 7.2 shows the three statistics calculated for Feigl and Zelen's data. Here we have ignored the dependence on white blood count and simply compared the survival times in the AG-positive and AG-negative groups. The value 7.95 for the score statistic based on the approximate log likelihood may be compared with the likelihood ratio test statistic from the same approximate log likelihood of $2(86.00 - 82.23) = 7.54$ from Table 7.1. The strong conservatism of the $\sum(D - E)^2/E$ approximation is largely due to the inclusion of the six AG-positive survival times that exceeded all the AG-negative survival times. These have no effect on the first two statistics, but inflate the variance approximation for the third. They would better have been excluded from the calculation of variance, yielding an approximate chi-squared of $6.703^2(1/17.703 + 1/9.297) = 7.37$.

The log rank test, in all its variants, has been derived as a test of the null hypothesis $\psi = 1$ under the proportional hazards model $h_1(t) = \psi h_0(t)$ or its discrete logistic analogue. For alternatives outside this class, it may have poor properties. Thus if $h_1(t) > h_0(t)$ for $t < t_0$, $h_1(t) < h_0(t)$ for $t > t_0$ the contributions to $U(0)$ from times τ with $\tau < t_0$

Table 7.2 Two-sample test for Feigl and Zelen's (1965) data (Table 1.2)

Time j	Group 0		Group 1		Total		$\dfrac{d_j r_{1j}}{r_j}$	$\dfrac{d_j r_{0j} r_{1j}}{r_j^2}$	$\dfrac{d_j r_{0j} r_{1j}(r_j - d_j)}{r_j^2(r_j - 1)}$
	r_{0j}	d_{0j}	r_{1j}	d_{1j}	r_j	d_j			
1	16	0	17	2	33	2	1.030	0.500	0.484
2	16	1	15	0	31	1	0.484	0.250	0.250
3	15	3	15	0	30	3	1.500	0.750	0.698
4	12	3	15	1	27	4	2.222	0.988	0.874
5	9	0	14	1	23	1	0.609	0.238	0.238
7	9	1	13	0	22	1	0.591	0.242	0.242
8	8	1	13	0	21	1	0.619	0.236	0.236
16	7	1	13	1	20	2	1.300	0.455	0.431
17	6	1	12	0	18	1	0.667	0.222	0.222
22	5	1	12	1	17	2	1.412	0.415	0.389
26	4	0	11	1	15	1	0.733	0.196	0.196
30	4	1	10	0	14	1	0.714	0.204	0.204
39	3	0	10	1	13	1	0.769	0.178	0.178
43	3	1	9	0	12	1	0.750	0.188	0.188
56	2	1	9	1	11	2	1.636	0.298	0.268
65	1	1	8	2	9	3	2.667	0.296	0.222
100	0	0	6	1	6	1	1.000	0	0
108	0	0	5	1	5	1	1.000	0	0
121	0	0	4	1	4	1	1.000	0	0
134	0	0	3	1	3	1	1.000	0	0
143	0	0	2	1	2	1	1.000	0	0
156	0	0	1	1	1	1	1.000	0	0
		16		17		33	23.703	5.654	5.319

$U(0) = 17 - 23.703 = -6.703$;

discrete chi-squared: $6.703^2/5.319 = 8.45$,

first approximate statistic: $6.703^2/5.654 = 7.95$.

will tend to be positive, those from failure times $\tau > t_0$ will tend to be negative and the result may be a nonsignificant test statistic, even if the two survival distributions have radically different shapes. The log rank test should be supplemented by cumulative hazard plots of the two distributions, and possibly also by analysis involving time-dependent covariates (Chapter 8) to protect against such disasters.

Chapter 8 will also discuss a variety of other two-sample test statistics, which differ from the log rank test in assigning greater weight to the earlier failure times, and less to the later times when few subjects remain alive.

The discussion at the end of Section 7.4 shows that, for uncensored data, the log rank test reduces to a test based on exponential order statistics. This aspect will be pursued later.

7.8 Estimation of baseline functions: goodness of fit

Attention in this chapter has primarily been focused on the regression parameters β rather than the baseline hazard function $h_0(t)$, the corresponding cumulative hazard

$$H_0(t) = \int_0^t h_0(u) \, du$$

and survivor function

$$\mathscr{F}_0(t) = \exp[-H_0(t)].$$

Indeed the conditioning argument completely eliminates these functions from the likelihood. If it is desired to express the conclusions of an analysis directly in terms of the survivor functions of particular subjects or groups of subjects, consideration of these baseline functions is essential. They also provide fairly straightforward graphical procedures for assessing goodness of fit.

If $h_0(t)$ is specified parametrically, say as $h_0(t, \phi)$, then ϕ may be estimated from the joint log likelihood $l(\beta, \phi)$. As in Chapter 6, a joint maximization in β and ϕ may be performed, or it may be more convenient to set $\beta = \hat{\beta}$, the conditional likelihood estimator, and maximize $l(\hat{\beta}, \phi)$ in ϕ. In the spirit of this chapter, it is more natural to seek nonparametric estimators.

A simply computed nonparametric estimator of $H_0(t)$ is obtained by noting that the total number of failures in $(0, t)$

$$D(t) = \sum_{\tau_j < t} d_j$$

has the same expectation as the total cumulative hazard

$$H(t) = \sum_{i=1}^{n} H_i(y_i)$$

up to time t for all study subjects. Here $y_i = t$ if $t_i > t$, so that the subject is still at risk at t, and otherwise equals his failure or censoring time. Thus,

$$H(t) = \int_0^t \sum_{j \in \mathcal{R}(u)} \psi(j) h_0(u) \, du,$$

suggesting the estimator

$$\hat{H}_0(t) = \sum_{\tau_j < t} \frac{d_j}{\sum_{l \in \mathcal{R}(\tau_j)} \hat{\psi}(l)},$$

where the $\hat{\psi}(l)$ are the estimated values of $\psi(l)$. Note that for a single homogeneous sample, this estimator reduces to the cumulative hazard estimator (4.14) of Chapter 4.

The baseline survivor function $\mathcal{F}_0(t)$ can be estimated by $\hat{\mathcal{F}}_0(t) = \exp[-\hat{H}_0(t)]$. Combining this estimator with the estimated multipliers $\hat{\psi}(i)$, we obtain the estimated values

$$\hat{H}_i(t) = \hat{\psi}(i)\hat{H}_0(t), \qquad \hat{\mathcal{F}}_i(t) = [\hat{\mathcal{F}}_0(t)]^{\hat{\psi}(i)}$$

for the hazard function and survivor function for subject i. With the aid of these estimators, graphical illustrations of the estimated effects of the explanatory variables on survival time can be given, for example by plotting a fixed percentile of the estimated survivor function $\hat{\mathcal{F}}_i(t)$ against the explanatory variables.

Another important use of these estimators is in assessing goodness of fit. If the survival times T_i were transformed by the true cumulative hazard functions $H_i(t)$, the resulting values, $H_i(T_i)$ would constitute a (censored) sample from the unit exponential distribution. The same result may be expected to hold approximately, when the T_i are transformed by the estimated functions $\hat{H}_i(t)$. The transformed $\hat{H}_i(T_i)$ may be called generalized residuals; see Section 6.6.

As an overall check on the model, the ordered values of $\hat{H}_i(T_i)$, say $\hat{H}_{(i)}$, may be plotted against their expected values for the unit exponential distribution. More usefully, separate plots may be made for subsets of the data defined by the explanatory variables. Such plots may indicate that additional explanatory variables or combinations of explanatory variables should be included in the model. Plots of empirical survivor functions from subsets of the data defined by values of the linear predictor may reveal departures from the proportional hazards model. The detailed sampling theory of these generalized residuals has not been investigated, however, and caution is needed in interpreting the results. Estimation of the multipliers ψ and the cumulative hazard function H_0 introduces a correlation between the residuals which may result in spuriously good fits to the theoretical form.

Bibliographic notes, 7

The proportional hazards model was first proposed by Cox (1972), who emphasized the log linear form for the multiplier ψ. The linear form has been used by Thomas (1981). Efron (1977) suggested the logistic form and Burridge (1981c) and Aranda-Ordaz (1980) have studied the more general families. Cox (1972) derived the likelihood (7.3) as a product of conditional probabilities. Kalbfleisch and Prentice (1973) pointed out the explicit interpretation in terms of the marginal likelihood of ranks.

Our exposition of the log linear hazards model and the discrete logistic model follows Cox (1972). See Howard (1972) and Gail et al. (1981) for details of the recursive computation scheme for the logistic log likelihood. The grouped marginal likelihood was derived by Kalbfleisch and Prentice (1973) and the approximations (7.18), (7.19) and (7.20) were suggested by, respectively, Peto (1972), Efron (1977) and Oakes (1981). Prentice and Gloeckler (1978) gave the unconditional analysis with a separate nuisance parameter for each grouping interval; see Exercise 7.5.

Mantel (1966) derived the log rank test for comparing survival distributions by analogy with Mantel and Haenszel's (1959) test (see also Cochran, 1954) for combining information from contingency tables. The derivation as a score test from the discrete logistic model is due to Cox (1972). Brown (1984) discusses variance estimators for the log rank test. See Peto et al. (1976, 1977) for an exposition of methods based on the approximate statistic $\chi^2 = \sum (D_i - E_i)^2 / E_i$.

Estimators for the hazard function were suggested by Cox (1972), Breslow (1972, 1974, 1975), Oakes (1972) and others. The notion of generalized residuals is due to Cox and Snell (1968); for the application to survival times, see Crowley and Hu (1977) and Kay (1977). Lagakos (1981) cautions their interpretation.

Further results and exercises, 7

7.1. Show that (7.4) can be derived as the sum of all probabilities (7.3) consistent with the observed pattern of failures and censoring. [Hint: Use mathematical induction, starting with the largest censored time.]

7.2. Suppose that there are two groups, with $n = 3$ observations per group and no censoring. For each of the 20 possible orderings of the failure times in the two groups, calculate the values of $U(0), I(0)$ and $\tilde{I}(0)$. Show that $|U(0)|$ is highly correlated with $I(0)$. Comment.

[Brown, 1984]

7.3. Show that if Peto's approximate likelihood is used, the second derivative of $\beta = 0$ overestimates the variance of the first derivative (for a given risk set) by a factor of $(r - 1)/(r - d)$.

7.4. Investigate the performance of an approximate likelihood, where each denominator takes the form

$$\left(\sum_{k \in \mathcal{R}_j} \psi(k) - \xi \sum_{k \in \mathcal{D}_j} \psi(k) \right)^d \qquad (0 < \xi < 1).$$

[Oakes, 1981]

7.5. Suppose that the continuous-time proportional hazards model (7.1) holds but that the observations are grouped into intervals $0 = a_0 < a_1 < \ldots < a_g$. Show that the 'grouped' hazards

$$h_j(z) = \operatorname{pr}(T < a_j | T \geqslant a_{j-1})$$

satisfy

$$\log[1 - h_j(z)] = \psi(z; \beta) \log[1 - h_j(0)].$$

For the case $\psi(z; \beta) = e^{\beta^{\mathrm{T}} z}$ derive the maximum likelihood estimator of β when the $h_j(0)$ are regarded as 'nuisance' parameters and investigate the large-sample theory ($n \to \infty$, g fixed).

[Prentice and Gloeckler, 1978]

7.6. A nearest-neighbour product-limit estimate is defined by first specifying a distance measure for the explanatory variables z. For each individual, a product-limit estimate of the survivor function is then computed by taking that individual plus its k nearest neighbours in terms of z. Suggest how, by taking the mean or median of the estimated survivor function, a graphical analysis can be made, at least when z is one-dimensional, and alternative models compared.

[Doksum and Yandell, 1983]

7.7. For the log rank scores q_i of (7.14), show that

$$\sum q_i^2 = d - \sum_{j=1}^{d} 1/r_j.$$

This gives a slightly simpler form for the permutation variance (7.15).

CHAPTER 8

Time-dependent covariates

8.1 Introduction

A time-dependent explanatory variable, called here a covariate, is one whose value for any given individual may change over time. This chapter considers the inclusion of such variables in the log linear proportional hazards model. After describing some applications, we discuss the partial likelihood estimation of the coefficients of time-dependent explanatory variables and compare the asymptotic efficiency of this method with that of fully parametric analysis of the same model. It is shown how a number of two-sample tests can be derived within the context of the log linear hazards model by the introduction of stochastic covariates. These tests have good properties against alternatives in which the ratio of hazards approaches unity with time, and are less sensitive than the log rank test to the tails of the distribution. A different derivation via the accelerated life model is also indicated. Finally, we indicate a technique for reducing the computational burden for fitting the log linear proportional hazards model, which also has wider applicability in industrial life-testing situations, where the values of relevant covariates may be determined only through destructive testing.

8.2 Some applications of time-dependent covariates

(i) Testing the validity of the proportional hazards model

The simple proportional hazards model for comparing two groups has

$$h_i(t) = \exp(\beta z_i)h_0(t),$$

where $z_i = 0$ or 1 for groups 0 and 1 respectively. The hazard ratio is e^β, which, of course, does not depend on t. As indicated in Chapter 5, by defining a time-dependent covariate

$$z_i(t) = \begin{cases} 0 & \text{group 0,} \\ \log[h_1(t)/h_0(t)] & \text{group 1,} \end{cases}$$

we can in principle allow arbitrary hazard functions in the two groups. More realistically, simple forms of departure from the proportional hazards model may be examined by suitable choices of components of $z_i(t)$. For example, if

$$z_{i1} = \begin{cases} 0 & \text{group 0,} \\ 1 & \text{group 1,} \end{cases} \qquad z_{i2} = \begin{cases} 0 & \text{group 0,} \\ t - t^* & \text{group 1,} \end{cases}$$

then the ratio of hazards becomes

$$\exp[\beta_1 + \beta_2(t - t^*)].$$

Here, t^*, an arbitrary constant chosen close to the mean survival time, is included to improve interpretability and to avoid numerical instabilities in the calculation. With this parameterization, e^{β_1} represents the ratio of the hazard functions in the two groups at t^*. A nonzero value of β_2 would imply an increasing ($\beta_2 > 0$) or decreasing ($\beta_2 < 0$) trend in the ratio of hazards with time.

A similar approach can be adopted with numerical covariates. For any component z_r, a time-dependent covariate of the form $z_r'(t) = (t - t^*)(z_r - z_r^*)$ represents an interaction effect between z_r and time on the ratio of hazards. It is often found that the predictive power of a covariate z_i decreases to zero as $t \to \infty$. This can be allowed for by the inclusion of a time-dependent covariate $e^{-\gamma t} z_i$, representing a transient effect as at (5.30).

Such time-dependent covariates are fixed as functions of time for each individual and are in no sense random quantities.

(ii) Measurements made during a study

Repeated determinations may be made during the course of a study of variables thought to be related to survival. For example, blood pressure will usually be measured at roughly regular intervals for each surviving patient, and the hazard function at a specified point in time may be influenced more by the subject's blood pressure at that time than by its value on entry to the study. Instead of a single value z_i of blood pressure, we would then have, for each patient i, a function $z_i(t)$

whose value at time t would be the level of the last reading before t. Because such a covariate may well be affected by 'treatments' applied to individuals, considerable care is needed in interpreting the result of fitting such models; see Section 5.1.

In epidemiological studies of mortality from occupational factors such as asbestos dust, the relevant effective cumulative exposure will often tend to increase during an individual's lifetime. As a crude approximation to the biologically relevant dose, it may be necessary simply to take years of exposure, or some weighted average of these. Sometimes it may be possible to assign approximate levels of exposure to particular jobs and thus derive a measure which takes into account concentration as well as duration of exposure. In either case, the relevant exposure at time t will be a function $z_i(t)$ for each subject i.

In industrial life-testing, it may be possible to estimate the wear or damage sustained by components while they are functioning, and this may be the most useful predictor of the instantaneous risk of failure.

The time-dependent covariates of these examples are random quantities. Their values on any study subject cannot usually be determined in advance. Despite this, and bearing in mind the interpretive difficulties mentioned in Chapter 5, it is usually sensible to condition on the actual realizations $z_i(t)$ of the random functions $Z_i(t)$.

(iii) Switching treatments

Time-dependent covariates can be used to model the effect of a subject transferring from control group to treatment group, or vice versa. This may happen if, for some reason, the administration of the treatment is delayed. An example would be the evaluation of intensive care units for the treatment of subjects who have just suffered a severe heart attack, with death as the outcome measure. The risk of death between the initial heart attack and admission to the unit may be substantial, and must be considered in any assessment of the effect of the unit on survival.

Supposing that the time of onset of the heart attack is known, so that it can serve as the time origin, then a simple model has

$$h_i(t) = h_0(t) \qquad (8.1)$$

for the hazard function of a subject t hours after his heart attack, if he has not by then been admitted to the unit, but

$$h_i(t) = e^\beta h_0(t) \tag{8.2}$$

if, by that time, he has been admitted. Thus the treatment, once it is applied, is assumed to multiply the hazard function by e^β. The effect on survival will be beneficial if $\beta < 0$, and harmful if $\beta > 0$. It must be stressed that estimation of β here would require the collection of data on a defined population of subjects who had suffered heart attacks and would have been admitted to the unit if they had survived long enough. Information on those who had been admitted would not in itself suffice. The minimal data required would include for each patient the time of the initial heart attack, the times of death and admission to the unit if these events occurred, and the total follow-up time for that patient.

Given this information, a covariate function $z_i(t)$ can be defined for each patient as

$$z_i(t) = \begin{cases} 0 & \text{if the patient has not yet been admitted at } t, \\ 1 & \text{if the patient has been admitted at } t, \end{cases}$$

and the model (8.1) and (8.2) becomes

$$h_i(t) = e^{\beta z_i(t)} h_0(t).$$

A much-studied example in similar vein concerns the evaluation of the Stanford heart transplant programme. Here there is a defined point of entry, the conference date, when the patient is judged suitable for a heart transplant. Dependence on the availability of a suitable donor heart means that the patient may not receive a new heart until weeks after the conference date, if at all. With time t measured from the conference date, the covariate function $z_i(t)$ takes the value zero or one according as patient i has not or has received a new heart by time t. This example will be discussed further in Section 8.9.

(iv) *Evolutionary covariates*

For certain purposes, it is useful to introduce a new class of random time-dependent covariates. Let $\mathcal{H}_t$ denote the history of failures, censoring and of any other random features of the problem, all up to time t. We shall call $Z(t)$ an evolutionary covariate if it is a function of $\mathcal{H}_t$ only. Thus, $Z(t)$ could be the size of the risk set at t, the number of failures before time t, or, in a comparison of two groups, the number of failures in one group before time t. The log linear hazards model with evolutionary covariates becomes

$$\lim_{\Delta \to 0} \frac{1}{\Delta} \mathrm{pr}\{i \text{ fails in } (t, t + \Delta) | i \in \mathscr{R}(t), \mathscr{H}_t\} = h_0(t) e^{\beta z_i(t)},$$

where $z_i(t) = z_i(t, \mathscr{H}_t)$ is the realized value of the evolutionary covariate at time t.

It is not permissible in mathematical arguments to condition on the entire realizations of the processes $Z_i(t)$, for their value at a specific time t_0, say, will carry information about the failures that occur in $t < t_0$. Nevertheless, we shall see in Section 8.4 that evolutionary covariates can legitimately be used, and this possibility extends the scope of the log linear hazards model.

Evolutionary covariates parallel the notion of evolutionary censoring introduced in Chapter 1, where the decision to censor a subject at t may depend on any event in $\mathscr{H}_t$.

One caveat should be noted. Often, as mentioned in Chapter 1, the time t is measured from different origins, in calendar time, for different subjects. Patients may enter clinical trials on different dates, with survival time measured for each patient from his own date of entry. If so, and if treatment assignment depends on previous failures, in calendar time, then the implied rearrangement of the failures may destroy the evolutionary properties of the covariates. The issues here have yet to be fully clarified. They are quite separate from the important practical question of possible secular trends in the dependence of failures on covariates, which can be handled by the inclusion of calendar time as an additional explanatory variable in the log linear hazards model, and by derived product variables to cover changes with time in any treatment effects present.

8.3 The likelihood function

Our derivation of the likelihood function for the parameter β in the log linear hazards model with time-dependent covariates closely parallels that given in the previous chapter for constant covariates. As before, let the d ordered observed failures occur at $\tau_1 < \tau_2 < \ldots < \tau_d$, and set $\mathscr{I}_j = i$ if subject i fails at τ_j. Let $\mathscr{R}(\tau_j) = \{i : t_i \geq \tau_j\}$ be the risk set at τ_j, and the history $\mathscr{H}_j$ be the collection of all failures and censorings before τ_j, and also the values of all covariates, including evolutionary covariates, up to and including time τ_j. The time τ_j itself is also included in $\mathcal{H}_j$, but the identity $\mathcal{I}_j$ of the subject which fails at τ_j is not. The conditional probability that $\mathcal{I}_j = i$ given $\mathcal{H}_j$ is then, simply,

$$p_j(i|\mathcal{H}_j) = \frac{\exp[\beta^T z_i(\tau_j)]}{\displaystyle\sum_{k \in \mathcal{R}(\tau_j)} \exp[\beta^T z_k(\tau_j)]}. \tag{8.3}$$

Inference about β is based on the product

$$\text{lik} = \prod_{j=1}^{d} p_j(i_j|\mathcal{H}_j).$$

To avoid double subscripts it is convenient to write, with a slight abuse of notation, z_{ki} for the covariate function z_k of the kth subject evaluated at the failure time t_i of the ith subject. Then we have

$$\text{lik} = \prod_{i \in \mathcal{D}} \frac{\exp(\beta^T z_{ii})}{\displaystyle\sum_{k \in \mathcal{R}_i} \exp(\beta^T z_{ki})}. \tag{8.4}$$

Justification for treating (8.4) as a likelihood function will be given in the next section.

As in Sections 7.3 and 7.4, we have that the log likelihood is

$$l = \sum_{i \in \mathcal{D}} \left[\beta^T z_{ii} - \log\left(\sum_{k \in \mathcal{R}_i} \exp(\beta^T z_{ki}) \right) \right]$$

with derivatives

$$U_r(\beta) = \sum_{i \in \mathcal{D}} [z_{iir} - A_{ir}(\beta)],$$

$$-I_{rs}(\beta) = -\sum_{i \in \mathcal{D}} C_{irs}(\beta),$$

where now

$$A_{ir}(\beta) = \frac{\displaystyle\sum_{k \in \mathcal{R}_i} z_{kir} \exp(\beta^T z_{ki})}{\displaystyle\sum_{k \in \mathcal{R}_i} \exp(\beta^T z_{ki})},$$

$$C_{irs}(\beta) = \frac{\displaystyle\sum_{k \in \mathcal{R}_i} z_{kir} z_{kis} \exp(\beta^T z_{ki})}{\displaystyle\sum_{k \in \mathcal{R}_i} \exp(\beta^T z_{ki})} - A_{ir}(\beta) A_{is}(\beta).$$

The subscripts r and s, as usual, indicate components of the vectors z.

Inference about β may proceed as before. Substantial computing may be involved if there are many failures and the dimensionality of β is at all large, as a three-way array of covariate values is used.

8.4 Partial likelihood

In Section 7.2, we showed that when the covariates do not depend on time, the product lik can be interpreted in terms of the marginal likelihood of ranks. With time-dependent covariates, this is no longer possible. However, each term $\mathcal{H}_j$ encompasses all previous terms τ_1, $\mathcal{H}_1, i_1, \tau_2, \mathcal{H}_2, i_2, \ldots, \tau_j$, that is, these quantities are all functions of $\mathcal{H}_j$. This allows three fundamental properties of the derivatives of the log likelihood to be derived quite easily. These properties, together with regularity conditions, allow the asymptotic properties of the log likelihood to be inferred.

The full likelihood of all the data may be written

$$
\text{lik}_f = \prod_{j=1}^{d} \left[g_j(\tau_j, \mathcal{H}_j | \mathcal{H}_{j-1}, i_{j-1}) \, p_j(i_j | \mathcal{H}_j) \right]
$$
$$
\times \, g_{d+1}(\mathcal{H}_\infty | \mathcal{H}_d, i_d).
$$

Here, g_j is the conditional joint density/distribution of the jth ordered failure time τ_j, of any censorings in (τ_{j-1}, τ_j) and of the evolution of any random covariate functions $\{Z_j(t): \tau_{j-1} \leqslant t \leqslant \tau_j\}$ throughout the interval (τ_{j-1}, τ_j). The last term is void (unity) unless some subjects remain in view after the final observed failure.

The likelihood lik is obtained from lik_f by omitting the terms g_j which correspond to the information provided by the gaps between successive failures. For this reason, it is called a partial likelihood. Now $p(i_j | \mathcal{H}_j)$ is a discrete distribution over the risk set $\mathcal{R}(\tau_j)$, and it depends on the parameter β. It follows, just as for any probability density or distribution, that

$$
E\left(\frac{\partial \log p_j(i_j | \mathcal{H}_j)}{\partial \beta_r} \middle| \mathcal{H}_j \right) = 0, \tag{8.5}
$$

$$
E\left(\frac{\partial^2 \log p_j(i_j | \mathcal{H}_j)}{\partial \beta_r \partial \beta_s} \middle| \mathcal{H}_j \right)
$$
$$
= -E\left(\frac{\partial \log p_j(i_j | \mathcal{H}_j)}{\partial \beta_r} \cdot \frac{\partial \log p_j(i_j | \mathcal{H}_j)}{\partial \beta_s} \middle| \mathcal{H}_j \right).
$$

All the expectations are taken conditional on the history $\mathcal{H}_j$. Since, however, these relations hold whatever the observed $\mathcal{H}_j$, they must also hold unconditionally. Thus,

$$E\left(\frac{\partial \log p_j(i_j|\mathcal{H}_j)}{\partial \beta_r}\right) = 0,$$

$$E\left(\frac{\partial^2 \log p_j(i_j|\mathcal{H}_j)}{\partial \beta_r \partial \beta_s}\right)$$

$$= -E\left(\frac{\partial \log p_j(i_j|\mathcal{H}_j)}{\partial \beta_r} \frac{\partial \log p_j(i_j|\mathcal{H}_j)}{\partial \beta_s}\right).$$

The properties of iterated expectations yield one further result. Since $\mathcal{H}_k$ and i_k are included in $\mathcal{H}_j$ if $k < j$,

$$E\left(\frac{\partial \log p_j(i_j|\mathcal{H}_j)}{\partial \beta_r} \frac{\partial \log p_k(i_k|\mathcal{H}_k)}{\partial \beta_s}\right)$$

$$= E\left[E\left(\frac{\partial \log p_j(i_j|\mathcal{H}_j)}{\partial \beta_r} \frac{\partial \log p_k(i_k|\mathcal{H}_k)}{\partial \beta_s}\bigg|\mathcal{H}_j\right)\right]$$

$$= E\left[\frac{\partial \log p_k(i_k|\mathcal{H}_k)}{\partial \beta_s} E\left(\frac{\partial \log p_j(i_j|\mathcal{H}_j)}{\partial \beta_r}\bigg|\mathcal{H}_j\right)\right]$$

$$= 0,$$

since the last term is identically zero in $\mathcal{H}_j$. Note that this argument holds if $r = s$, though not, of course, if $j = k$.

The standard proofs of the asymptotic normality of maximum likelihood estimates of a parameter θ for independent and identically distributed random variables X_j proceed by applying a central limit theorem to the sum

$$U(\theta) = \sum_{j=1}^{n} \frac{\partial \log p(x_j, \theta)}{\partial \theta}$$

whose terms are independent and identically distributed vectors, noting that $E[U(\theta)] = 0$ and that

$$E[U_r(\theta)U_s(\theta)] = -E[\partial U_r(\theta)/\partial \theta_s],$$

and expanding $U(\theta)$ in a Taylor series around the true parameter value. Some conditions on the third derivatives of the log likelihood are needed.

It has been shown that the increments of the derivative of the log partial likelihood, evaluated at the true parameter value, are uncorrelated. This is a much weaker result than full independence for all parameter values in a neighbourhood of the true value. Rigorous proofs of asymptotic normality of the estimators are, therefore, difficult and have been derived only in special cases. Nevertheless, it seems likely that the results hold in most circumstances of practical interest. Some restrictions are certainly needed to ensure that the total information from the sample does not remain bounded as the sample size increases, and that it is not dominated by a few failure times. Exercise 8.1 gives one simple instance where asymptotic normality does not hold.

The preceding development has ignored the fact that, when there is censoring, the number of terms d in the partial likelihood is itself a random variable. However, this does not introduce any major new ideas.

8.5   Efficiency comparisons with fully parametric models

If the underlying hazard $h_0(t)$ is known up to a fixed number of parameters, say to a parameter vector γ, the gap terms g_i may provide useful information about β. An important question concerns how much precision is lost by omitting these terms from the likelihood. We first consider further the form of the information matrix from the partial likelihood.

For each $t > 0$ define

$$A(\beta, t) = \frac{\sum\limits_{k \in \mathscr{R}(t)} z_k(t) \exp[\beta^T z_k(t)]}{\sum\limits_{k \in \mathscr{R}(t)} \exp[\beta^T z_k(t)]},$$

$$C(\beta, t) = \frac{\sum\limits_{k \in \mathscr{R}(t)} z_k(t) z_k^T(t) \exp[\beta^T z_k(t)]}{\sum\limits_{k \in \mathscr{R}(t)} \exp[\beta^T z_k(t)]} - A(\beta, t) A^T(\beta, t).$$

Thus $A(\beta, t)$ is the weighted average of the values of the covariate vector $z_k(t)$, taken over the risk set $\mathscr{R}(t)$, with weights proportional to the multipliers $\exp[\beta^T z_k(t)]$, and $C(\beta, t)$ is the corresponding covariance matrix.

Let $F_n(t)$ denote the improper empirical distribution of the observed failure times, so that $nF_n(t)$ counts the number of observed failures less than t. The first and second derivatives of the log likelihood may be written

$$U(\beta) = \sum_{i \in \mathcal{D}} z_{ii} - n \int_0^\infty A(\beta, t)\, dF_n(t), \qquad (8.6)$$

$$-I(\beta) = -n \int_0^\infty C(\beta, t)\, dF_n(t). \qquad (8.7)$$

Suppose now that, as $n \to \infty$, the random quantities $A(\beta, t)$ and $C(\beta, t)$ converge to deterministic limits $E(\beta, t)$ and $V(\beta, t)$. Then $E(\beta, t)$ can be interpreted as the expected value of the covariate vector evaluated on a subject observed to fail at t, and $V(\beta, t)$ is the corresponding covariance matrix. Suppose also, that as $n \to \infty$, the empirical distribution $F_n(t)$ converges to a limiting form $F(t)$. Integration with respect to $dF(t)$ is equivalent to taking an expectation with respect to the marginal distribution of the failure time, T. Thus, we may write, schematically,

$$n^{-1}U(\beta) \to E[Z(T)] - E\{E[Z(T)|T]\},$$

which, as we would expect, equals zero by the rule of iterated expectations. Also

$$n^{-1}I(\beta) \to E\{\mathrm{var}[Z(T)|T]\},$$

the expectation of a conditional variance.

The precision of maximum likelihood estimation from a fully parameterized model depends on the parameterization adopted for $h_0(t)$. The log linear form

$$h_0(t) = \exp[\alpha^T y(t)]$$

is a convenient choice. Here, α is a parameter vector of dimension p and $y(t)$ a specified set of covariate functions. Thus, $p = 1, y_1(t) = 1$ gives the exponential distribution, and $p = 2, y_1 = 1, y_2 = \log t$ gives a parameterization of the Weibull distribution.

The full hazard function for the ith subject is then

$$h_i(t) = \exp[\alpha^T y(t) + \beta^T z_i(t)].$$

To derive the log likelihood $l_f(\alpha, \beta)$ of the full data in α and β, we make the important assumption that the covariate functions $z_i(t)$, although they may be time-dependent, are deterministic, or if not, that their distributions do not depend on the parameters α and β. Note that we do not exclude evolutionary covariates whose conditional distributions satisfy the same assumptions. This assumption, that the values of the covariates in themselves carry no information about the parameters of interest, is the mathematical counterpart of the point made in Chapter 5, concerning the possible role of time-dependent covariates as intermediate outcomes. A similar assumption is made about the censoring times. Subject to these assumptions,

$$l_f(\alpha, \beta) = \sum_{i \in \mathscr{D}} [\alpha^T y(t_i) + \beta^T z_i(t_i)]$$

$$- \int_0^\infty \sum_{k \in \mathscr{R}(t)} \exp[\alpha^T y(t) + \beta^T z_k(t)] dt. \qquad (8.8)$$

Let $p = \dim(\alpha)$, $q = \dim(\beta)$. Then the $(p+q) \times (p+q)$ joint information matrix in α and β is the expectation of the matrix $- \partial^2 l_f / \partial \alpha \partial \beta$. It is easily shown to equal the expectation of the matrix

$$S = \begin{pmatrix} S_{\alpha\alpha} & S_{\alpha\beta} \\ S_{\alpha\beta}^T & S_{\beta\beta} \end{pmatrix},$$

where

$$S_{\alpha\alpha} = \sum_{i \in \mathscr{D}} y(t_i) y(t_i)^T,$$

$$S_{\beta\beta} = \sum_{i \in \mathscr{D}} z_i(t_i) z_i(t_i)^T,$$

$$S_{\alpha\beta} = \sum_{i \in \mathscr{D}} y(t_i) z_i(t_i)^T.$$

The inverse S^{-1} of S has as its $q \times q$ bottom right-hand submatrix $\bar{S}_{\beta\beta}$, where

$$\bar{S}_{\beta\beta}^{-1} = S_{\beta\beta} - S_{\alpha\beta}^T S_{\alpha\alpha}^{-1} S_{\alpha\beta}.$$

This estimates the marginal information $I_f(\beta)$ about β provided by the full log likelihood $l_f(\alpha, \beta)$ (cf. Section 3.3).

Some insight into these formulae may be gained from the special case $p = 1$, $y_1(t) \equiv 1$, $q = 1$, corresponding to $h_0(t) = e^\alpha$ and a single scalar covariate. For then

$$\bar{S}_{\beta\beta}^{-1} = \sum_{i \in \mathscr{D}} z_{ii}^2 - \left(\sum_{i \in \mathscr{D}} z_{ii} \right)^2 \bigg/ d.$$

In the notation used earlier,

$$n^{-1}I_f(\beta) \to \text{var}[Z(T)],$$

the unconditional variance of the covariate $Z(T)$ evaluated at the failure time T, where again, T has the improper distribution $F(t)$. The relation

$$\text{var}[Z(T)] = E\{\text{var}[Z(T)|T]\} + \text{var}\{E[Z(T)|T]\}$$

shows that the components of the normalized information about β from $l_f(\alpha,\beta)$ contained in the partial likelihood $l(\beta)$ and the gap terms $\{g_j\}$ are essentially the within- and between-group components of var$[Z(T)]$, with groups defined by the failure time, T. The partial likelihood will have high asymptotic efficiency relative to the full likelihood if the ratio of the between-group component to the within-group component is small. This condition will usually hold unless either

   (i)   the parameter β is far from zero;
  (ii)   censoring is strongly dependent on $Z(T)$; or
 (iii)   there are strong time trends in the covariates.

Full asymptotic efficiency for the partial likelihood requires that $E[Z(T)|T]$ have zero variance, so that it is a constant. This will hold for example, if the $z_i(t)$ are not time-dependent but constant, as in Chapter 7, if $\beta = 0$, and if the censoring is independent of z.

Extensions of these results to cases where $p > 1$ and or $q > 1$ are fairly straightforward. Thus, if $p > 1$ and $q = 1$, the decomposition of var $[Z(T)]$ is replaced by a decomposition of its variance around the multiple regression of $E[Z(T)|T]$ on $y(T)$. Full asymptotic efficiency will hold if $E[Z(T)|T]$ can be expressed as an exact linear combination of the $y_r(T)$.

These results are only asymptotic. For finite samples, the loss in precision from using the partial likelihood is rather greater.

8.6   Two-sample tests

In the absence of ties, the log rank statistic introduced in the previous chapter for comparing two distributions takes the form

$$U(0) = \sum_{j \in \mathscr{D}} (d_{1j} - A_j), \tag{8.9}$$

where $d_{1j} = 1$ or $d_{1j} = 0$ according as the failure at τ_j occurs in the first or second group, and $A_j = r_{1j}/r_j$ is the proportion of the total risk set $\mathcal{R}_j$ at τ_j which belongs to the first group. This test was derived by evaluating at $\beta = 0$ the derivative of the log partial likelihood from the proportional hazards model

$$h_i(t) = \exp(\beta z_i)h_0(t),$$

where $z_i = 1$ and $z_i = 0$ for the two groups.

Although the log rank test performs well against alternative hypotheses within the proportional hazards class, for alternatives outside this class it can be very sensitive to the tails of the distribution, that is, to failures that occur when very few subjects remain alive. To avoid this difficulty, a weighted sum can be formed,

$$U(0, w) = \sum_{j \in \mathcal{D}} w_j(d_{1j} - A_j). \tag{8.10}$$

with the early failure times receiving the greater weight.

The use of evolutionary covariates allows such tests to be derived and variance estimates calculated within the log linear hazards model, with the important proviso that the weight w_j may depend on the history of failures and censorings up to τ_j, but must not depend on what happens after τ_j. Define

$$z_i(t) = \begin{cases} w(t) & i \in \text{group } 1, \\ 0 & i \in \text{group } 0, \end{cases}$$

where $w(t) = w(t, \mathcal{H}_t)$ is chosen so that $w_j = w(\tau_j)$. It is easily seen that the first derivative of the log likelihood is just $U(0, w)$ and the second derivative, which gives the estimated variance of $U(0, w)$, is

$$I(0, w) = \sum_{j \in \mathcal{D}} w_j^2 A_j(1 - A_j).$$

Particular choices that have been suggested for w_j include

$w_j = r_j,$	(Gehan, 1965);
$w_j = n\hat{\mathscr{F}}(\tau_j),$	(Prentice, 1978); (8.11)
$w_j(\kappa) = n[\hat{\mathscr{F}}(\tau_j)]^\kappa \quad (0 \leqslant \kappa \leqslant 1),$	(Harrington and Fleming, 1982).

Here, $\hat{\mathscr{F}}(\tau_j)$ is an estimator of the combined survivor function of the two groups, usually taken in this context as

$$\hat{\mathscr{F}}_j = \prod_{k=1}^{j} \left(\frac{r_k}{r_k + 1} \right).$$

The first two tests are usually known as 'generalized Wilcoxon tests' because they reduce, in the absence of censoring, to the usual

Wilcoxon statistic. In current thinking, Prentice's version of the weights is preferred to Gehan's, because the latter are heavily dependent on the censoring pattern. The third suggestion is an attempt to compromise between the log rank test ($\kappa = 0$) and the generalized Wilcoxon test ($\kappa = 1$).

Other suggestions have been made. An early one of Efron was to consider

$$U = - \int_0^\infty \hat{\mathscr{F}}_2(t)\, d\hat{\mathscr{F}}_1(t),$$

where $\hat{\mathscr{F}}_1$ and $\hat{\mathscr{F}}_2$ are the separate product-limit estimators of the survivor functions in the two groups. This statistic is of interest because, unlike the others mentioned, it gives a consistent estimator of the probability

$$\mathrm{pr}(T_2 > T_1) = - \int \mathscr{F}_2(t)\, d\mathscr{F}_1(t), \tag{8.12}$$

whatever the underlying survivor functions $\mathscr{F}_1$ and $\mathscr{F}_2$, and for rather general censoring patterns. However, in the presence of heavy censoring, it can be unstable in the tails. Efron's test can also be obtained by setting

$$w(t) = \frac{\hat{\mathscr{F}}_1(t)\,\hat{\mathscr{F}}_2(t)\, r(t)}{r_1(t)\, r_2(t)}. \tag{8.13}$$

8.7 Rank tests and the accelerated life model

Suppose that the observations follow an accelerated life model of Section 6.3, but with scalar z, so that

$$Y_i = \log T_i = \mu - \beta z_i + \varepsilon_i = - \beta z_i + e_i, \tag{8.14}$$

where z_i is the value of the covariate on the ith subject and $e_i = \mu + \varepsilon_i$ has a known distribution with density f and survivor function $\mathscr{F}$. As before, let $\tau_1 < \ldots < \tau_d$ denote the ordered observed failure times ($\tau_0 = 0, \tau_{d+1} = \infty$). Let $\mathscr{M}_j (j = 0, 1, \ldots, k)$ be the set of m_j subjects censored between τ_j and τ_{j+1}. Consider the likelihood of the identities $\mathscr{I} = \{\mathscr{I}_j, \mathscr{M}_j\}$. In the absence of knowledge of the censoring mechanism, this cannot be obtained exactly, but to a close approximation it may be derived as follows.

Writing $y_j = \log \tau_j$ and $z_{(j)} = z_{(\mathscr{I}_j)}$ for the value of the covariate on the subject observed to fail at τ_j, we see that this subject contributes a term $f(y_j + \beta z_{(j)})$ to the likelihood in $(y, \mathscr{I})$. If it is assumed that the m_j subjects in $\mathscr{M}_j$ are observed to survive only beyond τ_j, then, together, they contribute a term

$$\prod_{i \in \mathscr{M}_j} \mathscr{F}(y_j + \beta z_i).$$

The total likelihood in β is, therefore,

$$E(\beta; \mathscr{I}, y) = \prod_{j=1}^{d} f(y_j + \beta z_{(j)}) \prod_{i \in \mathscr{M}_j} \mathscr{F}(y_j + \beta z_i). \qquad (8.15)$$

To obtain the likelihood in $\mathscr{I}$, we must integrate out with respect to the log failure times y over the permissible region $\Gamma(d) = y_1 < y_2 < \ldots < y_d$. Thus,

$$\text{lik}(\beta; \mathscr{I}) = \int_{\Gamma(d)} E(\beta; \mathscr{I}, y) dy. \qquad (8.16)$$

In general, the value of this integral will depend on the form of f, though it is the same for all members of a location family. The value at $\beta = 0$ does not depend on f; it equals

$$\text{lik}(0; \mathscr{I}) = \int_{\Gamma'(d)} \prod_{j=1}^{d} (1 - u_j)^{m_j} du = \prod_{j=1}^{d} 1/r_j,$$

where

$$r_j = \sum_{k=j}^{d} (1 + m_k)$$

is the size of the risk set at τ_j and $\Gamma'(d)$ is the transformed region $0 < u_1 < \ldots < u_d < 1$. This result is easily proved by mathematical induction.

Locally most powerful rank tests of the hypothesis $\beta = 0$ are obtained from the derivative of the log likelihood in β, evaluated at $\beta = 0$. Differentiation of $E(\beta; \mathscr{I}, y)$ with respect to β gives a sum of n terms, one per observation, each being $E(\beta; \mathscr{I}, y)$ itself, multiplied by either

$$z_{(j)} f'(y_j + \beta z_{(j)})/f(y_j + \beta z_{(j)})$$

or by

$$-z_i f(y_j + \beta z_i)/\mathscr{F}(y_j + \beta z_i),$$

as appropriate. Evaluation at $\beta = 0$ gives

$$\left[\frac{\partial \log \text{lik}(\beta; \mathcal{I})}{\partial \beta}\right]_{\beta=0} = \frac{1}{\text{lik}(0; \mathcal{I})} \int_{\Gamma(d)} \left[\frac{\partial E(\beta; \mathcal{I}, y)}{\partial \beta}\right]_{\beta=0} dy$$

$$= \sum_{j=1}^{d} \left(q_j^{(u)} z_{(j)} + \sum_{i \in \mathcal{M}_j} q_i^{(c)} z_i\right), \qquad (8.17)$$

where

$$q_j^{(u)} = \int_{\Gamma(d)} \frac{f'(y_j)}{f(y_j)} \prod_{k=1}^{d} r_k E(0; \mathcal{I}, y) \, dy,$$

$$q_j^{(c)} = \int_{\Gamma(d)} -\frac{f(y_j)}{\mathcal{F}(y_j)} \prod_{k=1}^{d} r_k E(0; \mathcal{I}, y) \, dy.$$

The transformation $u_k = F(y_k)$ allows the $q_j^{(u)}$ and $q_j^{(c)}$ to be written as

$$q_j^{(u)} = \int_{\Gamma'(d)} \psi(u_j) \prod_{k=1}^{d} r_k (1 - u_k)^{m_k} \, du, \qquad (8.18)$$

$$q_j^{(c)} = \int_{\Gamma(d)} \Psi(u_j) \prod_{k=1}^{d} r_k (1 - u_k)^{m_k} \, du, \qquad (8.19)$$

where

$$\psi(u) = \frac{f'[G(u)]}{f[G(u)]}, \qquad \Psi(u) = -\frac{f[G(u)]}{(1-u)} \qquad (8.20)$$

and $G(u)$ is the inverse function to $F(u)$. Explicit evaluation of the $q_j^{(u)}$ and $q_j^{(c)}$ is straightforward only for certain distributions. For the logistic density,

$$f(y) = e^y/(1 + e^y)^2$$

we have, easily, $\psi(u) = 1 - 2u, \Psi(u) = -u$. We obtain

$$q_j^{(u)} = 2 \prod_{k=1}^{j} \left(\frac{r_k}{r_k + 1}\right) - 1, \qquad q_j^{(c)} = \prod_{k=1}^{j} \left(\frac{r_k}{r_k + 1}\right) - 1. \qquad (8.21)$$

This is in fact the original derivation of Prentice's generalized

Wilcoxon test considered in Section 8.6. Exercise 8.5 demonstrates the equivalence of the two forms.

For the extreme value distribution, $f(y) = \exp(y - e^y)$, a change in location parameter $y \to y - \alpha$ corresponds to multiplication of the hazard function by e^α. It is therefore not surprising to find that this density yields the log rank test. We find

$$\psi(u) = \log(1 - u) + 1, \qquad \Psi(u) = \log(1 - u),$$

$$q_j^{(u)} = 1 - \sum_{k=1}^{j} \frac{1}{r_k}, \qquad q_j^{(c)} = - \sum_{k=1}^{j} \frac{1}{r_k}, \qquad (8.22)$$

giving (7.14).

8.8 Sampling the risk set

The partial likelihood (8.4) involves the value of every explanatory variable evaluated at each failure time for each subject surviving to that time. Determination of these values may involve considerable work. In some industrial reliability settings, where the variable represents the cumulative wear or damage sustained by a component, this may be determined only by destructive examination, which effectively censors the lifetime at that point. Even if determination of the values of the explanatory variables is straightforward, maximization of (8.4) can be computationally expensive.

From (8.4) the information about β contributed by a single risk set does not generally increase by virtue of its size. This suggests that considerable economy may be achieved at little cost in the precision of the estimator by comparing each failure with only a random sample of the corresponding survivors, a sample of size m, say.

The method may be justified as follows. Consider a particular risk set $\mathcal{R}(t) = \mathcal{R}$ of size r. Conditional on the history $\mathcal{H}_t$ up to t, a subset $\mathcal{S}$ of $\mathcal{R}$, of size $m + 1$, is selected in two steps. First, subject $i \in \mathcal{R}$ is selected with probability proportional to $\theta_i = \exp[\beta^T z_i(t)]$. This subject is the failure. Secondly, random sampling without replacement is used to select the remaining m elements of $\mathcal{S}$ from the $r - 1$ remaining elements of $\mathcal{R}$. These are the survivors in $\mathcal{S}$.

It is easily seen that the (conditional) probability of the subset $\mathcal{S}$, i.e. in the sample space consisting of all possible subsets of $\mathcal{R}$ of size $m + 1$, is proportional to $\sum_{j \in \mathcal{S}} \theta_j$, and that the conditional probability that i is the failure, given the composition of $\mathcal{S}$, is $\theta_i / \sum_{\mathcal{S}} \theta_j$. Thus, each term $p_j(i_j | \mathcal{H}_j)$ in lik can be expressed as the product of the conditional probability of $\mathscr{S}$ given $\mathscr{R}(t)$ and $\mathscr{H}_t$, with $t = \tau_j$, and the conditional probability of $i = i_j$ given $\mathscr{S}$. The product of the latter terms is itself a partial likelihood.

Insight into the loss of precision to be expected from discarding some of the (possible) data in this way is provided by the standard formula for the variance of a difference in means of independent samples,

$$\text{var}(\bar{x}_1 - \bar{x}_2) = \sigma^2(1/n_1 + 1/n_2).$$

If $n_1 = 1$, the gain in precision from increasing n_2 beyond 5 or 6 is small. In the more complex application considered here, it appears from simulation studies that 8 or 10 controls per case usually suffice to recover most (about 80%) of the information from the full partial likelihood.

8.9 An example

As an illustration of the use of time-dependent covariates we consider the survival of patients accepted into the Stanford heart transplant programme. The data, kindly made available by Dr Rupert Miller, are given in Table 8.1 and cover the period from the start of the programme in 1967, to February 1980. By this time a total of 249 patients had been accepted for transplantation, but only 184 patients had actually received transplants. Of the remaining 65 patients all but eight died awaiting transplantation. By contrast, 65 of the transplanted patients were still alive at the closing date.

The design of the programme is crucial to a proper understanding of the data. Patients are accepted into the programme when judged by the physicians to be suitable candidates for transplantation. When a donor heart becomes available, medical judgement is used to select the candidate who should receive it. For transplant patients both the waiting time, from the date of acceptance to the date of transplant, and post-transplant survival time are observed, together with an indicator of the final vital status. For nontransplant patients, the survival time from the date of acceptance and the vital status indicator are available.

Early evaluations of the programme in the medical literature compared the survival times of the nontransplant patients with those of the transplant patients, as if they had been allocated into the two groups on entry. As was indicated by Gail (1972) this procedure is

Table 8.1   *Survival times (days) of patients accepted into the Stanford heart
transplant programme. Follow up to February* 1980

Waiting time	Transplant indicator*	Survival post-transplant	Total survival	Final status†
49	2			1
5	2			1
0	1	15	15	1
35	1	3	38	1
17	2			1
2	2			1
50	1	623	673	1
39	2			1
84	2			1
11	1	46	57	1
25	1	126	151	1
7	2			1
16	1	64	80	1
36	1	1350	1386	1
0	2			1
27	1	279	306	1
35	2			1
19	1	23	42	1
36	2			1
17	1	10	27	1
7	1	1024	1031	1
11	1	42	53	1
2	1	730	732	1
82	1	136	218	1
24	1	1961	1985	1
112	2			2
262	2			1
70	1	1	71	1
34	2			1
15	1	836	851	1
15	2			1
16	1	60	76	1
50	1	1996	2046	1
22	1	3694	3716	2
11	2			1
45	1	49	94	1
18	1	47	65	1

Table 8.1 (*contd.*)

Waiting time	Transplant indicator*	Survival post-transplant	Total survival	Final status†
4	1	0	4	1
1	1	51	52	1
40	1	2878	2918	1
57	1	3410	3467	1
2	2			1
1	2			1
39	2			1
0	1	44	44	1
1	1	994	995	1
20	1	51	71	1
8	2			1
35	1	1478	1513	1
82	1	897	979	1
31	1	254	285	1
101	2			1
40	1	148	188	1
2	2			1
9	1	51	60	1
66	1	3021	3087	2
148	2			1
20	1	323	343	1
77	1	2984	3061	2
2	1	66	68	1
1	2			1
57	2			2
26	1	2723	2749	2
32	1	550	582	1
11	1	66	77	1
31	2			1
56	1	227	283	1
2	1	65	67	1
9	1	2805	2814	2
4	1	25	29	1
30	1	2734	2764	2
3	1	631	634	1
26	1	63	89	1
4	1	12	16	1

Table 8.1 (*contd.*)

Waiting time	Transplant indicator*	Survival post-transplant	Total survival	Final status†
1	2			1
45	1	2474	2519	1
20	2			1
209	1	547	756	1
66	1	29	95	1
25	1	1384	1409	1
5	1	544	549	1
31	1	48	79	1
36	1	297	333	1
4	2			1
7	1	1318	1325	1
59	1	50	109	1
30	1	1352	1382	1
138	1	68	206	1
159	1	26	185	1
340	2			1
309	1	146	455	1
27	1	431	458	1
4	1	161	165	1
1	1	14	15	1
12	1	2313	2325	2
20	1	1634	1654	1
95	1	48	143	1
20	2			1
37	1	2127	2164	1
56	1	263	319	1
50	1	2106	2156	2
70	2			1
1	1	293	294	1
5	1	2025	2030	2
29	1	2000	2029	2
1	1	2006	2007	2
1	1	1995	1996	2
10	1	1945	1955	2
6	1	65	71	1
2	1	731	733	1
40	1	1866	1906	2
18	1	538	556	1

Table 8.1 (contd.)

Waiting time	Transplant indicator*	Survival post-transplant	Total survival	Final status†
0	1	1846	1846	2
26	1	68	94	1
19	1	1778	1797	2
68	1	928	996	1
55	2			1
11	1	1722	1733	2
1	1	1718	1719	2
30	1	22	52	1
29	1	7	36	1
25	1	40	65	1
47	1	1612	1659	2
46	1	25	71	1
1	1	1638	1639	2
59	1	1547	1606	2
15	1	1534	1549	1
70	2			1
32	1	1271	1303	1
63	2			2
11	1	44	55	1
52	1	1232	1284	1
4	1	1247	1251	1
10	1	191	201	1
42	1	1393	1435	2
1	2			1
35	1	1202	1237	1
51	1	274	325	1
34	1	1373	1407	2
3	2			1
7	1	1378	1385	2
6	1	31	37	1
14	2			1
46	1	381	427	1
16	1	1341	1357	2
70	1	1262	1332	2
3	1	42	45	1
27	1	1261	1288	2
17	1	47	64	1
11	1	1264	1275	2

Table 8.1 (*contd.*)

Waiting time	Transplant indicator*	Survival post-transplant	Total survival	Final status†
82	1	48	130	1
202	1	30	232	1
86	1	1150	1236	1
70	1	626	696	1
38	1	1193	1231	2
71	1	45	116	1
43	2			1
63	1	1107	1170	2
129	1	1040	1169	1
69	2			1
12	1	1116	1128	2
25	1	1102	1127	2
39	1	195	234	1
13	2			1
53	2			1
36	1	993	1029	2
59	1	950	1009	2
45	1	121	166	1
4	1	729	733	1
35	1	202	237	1
48	1	841	889	2
20	1	1	21	1
88	1	752	840	1
0	1	834	834	2
1	1	265	266	1
121	1	132	253	1
76	1	738	814	2
26	1	86	112	1
10	1	328	338	1
2	1	793	795	2
10	1	781	791	2
86	1	663	749	2
33	2			1
35	2			2
30	1	221	251	1
75	1	90	165	1
9	1	660	669	2
79	2			1

Table 8.1 (*contd.*)

Waiting time	Transplant indicator*	Survival post-transplant	Total survival	Final status†
106	1	36	142	1
36	2			1
12	1	618	630	2
9	2			1
29	2			1
14	1	619	633	2
17	2			1
5	1	576	581	2
26	1	548	574	2
1	1	563	564	2
12	1	549	561	2
32	1	169	201	1
33	1	122	155	1
19	1	534	553	2
8	1	541	549	2
16	2			1
18	2			1
62	1	464	526	2
2	2			1
82	1	10	92	1
1	2			1
70	1	136	206	1
167	1	322	489	2
52	1	5	57	1
15	1	382	397	1
9	1	468	477	2
63	1	406	469	2
15	2			1
2	1	391	393	2
13	2			1
11	1	374	385	2
92	1	291	383	1
17	1	50	67	1
36	1	139	175	1
117	1	145	262	1
51	1	146	197	1
18	2			1
89	1	22	111	1

Table 8.1 (*contd.*)

Waiting time	Transplant indicator*	Survival post- transplant	Total survival	Final status†
223	2			2
59	1	231	290	1
65	1	188	253	1
82	1	149	231	2
27	1	176	203	2
192	2			2
67	1	119	186	2
18	2			1
176	2			2
9	1	138	147	1
11	2			1
146	2			2
125	2			2
15	1	107	122	2
23	1	98	121	2
31	2			1
30	1	89	119	2
22	1	56	78	2
24	2			1
10	1	60	70	2
25	1	2	27	2
14	2			2
12	1	1	13	2

* 1 = transplanted, 2 = not transplanted
† 1 = dead, 2 = alive

inappropriate here, as only those candidates who survive long enough to receive a heart will get one. The allocation procedure is heavily biased towards those patients with longer survival. Nor is the difficulty overcome by comparing the post-transplant survival of the transplant patients with the survival of the nontransplant patients. For, as was indicated in Chapter 1, it is important to choose as time origin a point where all individuals can be regarded as comparable. Patients who have already survived the critical few days or weeks from acceptance to transplant are in no sense comparable to those who are first embarking on this period.

A better procedure, indicated in Chapter 5 and Section 8.1, is to define a time-dependent indicator variable which takes the value zero or one at time t (measured from date of acceptance) according as the patient has not or has received a new heart by that time. Also since it might be thought that the few days immediately after transplantation might be the most hazardous period, for a transplant patient it is of interest to incorporate a transient effect, as suggested in Section 5.7. This leads to a model

$$h_i(t) = h_0(t),$$

for a patient who has not received a new heart by time t, and

$$h_i(t) = \psi_i(t)h_0(t),$$

where

$$\psi_i(t) = \exp(\beta_1 + \beta_2 e^{-\gamma v}),$$

for a patient who received a new heart at time $t - v$.

In view of the very small number of waiting times that exceed 6 months, analysis was limited to the first 183 days after acceptance. Results of fitting the model are shown in Table 8.2. The log likelihood

Table 8.2   *Stanford heart transplant data. Fitting of transient effect*

Model	Parameter value	Maximized log likelihood
Null	$\gamma = \quad .$ $\beta_1 = \quad 0$ $\beta_2 = \quad 0$	-555.66
Constant effect	$\gamma = \quad .$ $\beta_1 = -0.258$ $\beta_2 = \quad 0$	-555.07
Local maximum	$\gamma = \quad 0.715$ $\beta_1 = -0.287$ $\beta_2 = \quad 0.413$	-554.96
Local maximum	$\gamma = \quad 0.0563$ $\beta_1 = -0.159$ $\beta_2 = -0.285$	-554.96
$\gamma \to 0$	$\tilde{\beta}_1 = -0.128$ $\tilde{\beta}_2 = -0.00399$	-554.76

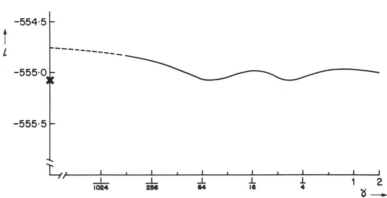

Fig. 8.1. Stanford heart transplant data. Fitting of model with transient effect containing term $e^{-\gamma t}$. Log likelihood for fixed γ maximized over other parameters. Model with no transient term, $\times$.

has a local maximum at $\gamma = 0.72, \beta_1 = -0.29, \beta_2 = 0.41$, corresponding to a slight beneficial long-term effect of transplantation coupled with a very rapidly decaying harmful transient effect. However, as is clear from the likelihood profile in γ plotted in Fig. 8.1, this is not a global maximum. There are several local maxima including one at 0.056, and a better fit is obtained by letting $\gamma \to 0$, so that $\psi_i(t)$ collapses to

$$\psi_i(t) = \tilde{\beta}_1 - \tilde{\beta}_2 v,$$

where $\tilde{\beta}_1 = \beta_1 + \beta_2, \tilde{\beta}_2 = \gamma\beta_2$.

None of the effects approaches statistical significance, but there is some slight evidence of a benefit from transplantation. No transient effects can be identified, perhaps because the major risk in transplantation is not from the surgery itself but from subsequent rejection or infection. Even these very tentative conclusions must be further cautioned. The analysis has ignored other explanatory variables, among which the age at acceptance is certainly important. Selection criteria for patients and the nature of the medical procedures followed have changed substantially over the period covered by these data. Most importantly, the validity of this analysis depends on there being no systematic tendency to select either the healthier or the less healthy patients for transplantation from the potential recipients at the time a donor heart becomes available. As no randomization was employed, this assumption is problematic.

Bibliographic notes, 8

Inclusion of time-dependent covariates as a device for testing the validity of the proportional hazards model was suggested by Cox (1972). The survival of patients in the Stanford heart transplant programme has been extensively analysed; see, for example, Gail (1972), Mantel and Byar (1974), Turnbull *et al.* (1974), Crowley and Hu (1977), Miller and Halpern (1982) and Aitkin *et al.* (1983), although the particular analysis given here appears new.

For another application of time-dependent covariates, see Farewell (1979). The use of evolutionary covariates was suggested by Lustbader (1980) and Oakes (1981).

The theory of Sections 8.3 and 8.4 is due to Cox (1972, 1975). The efficiency results of Section 8.5 were developed by Efron (1977) and Oakes (1977). For more specific results, including some finite sample calculations, see Kalbfleisch (1974), Kay (1979) and Kalbfleisch and McIntosh (1977). For asymptotic theory, see Liu and Crowley (1978), Tsiatis (1981), Sen (1981), Andersen and Gill (1982) and Self and Prentice (1982).

Wilcoxon tests for censored data were introduced by Gehan (1965) and Gilbert (1962); for a very good discussion, see Miller (1981, Chapter 4). Breslow (1970) derived an extended Kruskal–Wallis test for comparing several groups, along similar lines. Prentice (1978) suggested weights proportional to the Kaplan–Meier estimator. See Harrington and Fleming (1982) and Efron (1967) for the other variants mentioned. Gill (1980) gives a careful treatment of large-sample properties, using martingale theory.

The theory of Section 8.7 is due to Peto and Peto (1972) and Prentice (1978). See Mehrotra *et al.* (1982) for the connection between the weighted log rank and accelerated life derivations of the tests. Andersen *et al.* (1982) give a comprehensive review, emphasizing the martingale approach to the mathematical theory. Liddell *et al.* (1977) appear to have first advocated sampling from the risk set in the context of the log linear hazards model. For comments on the relative efficiency of this procedure, see Breslow and Crowley (1981) and Breslow *et al.* (1983).

Further results and exercises, 8

8.1. Suppose that from a sample originally of size n, all except m subjects are censored immediately after the first failure. The m

subjects are a random sample from the $n - 1$ survivors. Show that if $n \to \infty$, with m fixed, the total information about β will in general remain bounded.

Investigate the behaviour of the log rank test, and of Gehan and Prentice's versions of the Wilcoxon test, in these circumstances.

[Prentice and Marek, 1979]

8.2. Suppose that a subject with hazard function $h(t) = \exp[\alpha^T y(t)]$ is observed either to failure T or to a prespecified censoring time c, whichever occurs first. Show that the log likelihood in the vector α has a matrix of negative second derivatives whose (r,s)th element has the same expectation as a random variable which equals $y_r(T) y_s(T)$ if $T < c$ and otherwise equals zero.

8.3. Show by direct integration with respect to $u_{j+1}, u_{j+2}, \ldots, u_d$ that the $q_j^{(u)}$ and $q_j^{(c)}$ given by (8.18) and (8.19) may be written

$$q_j^{(u)} = \int_{\Gamma(j)} \prod_{k=1}^{j-1} [r_k(1 - u_k)]^{m_k}$$
$$\times [\psi(u_j) r_j(1 - u_j)^{r_j - 1}] \, du_1 du_2 \ldots du_j,$$

$$q_j^{(c)} = \int_{\Gamma(j)} \prod_{k=1}^{j-1} [r_k(1 - u_k)]^{m_k}$$
$$\times [\Psi(u_j) r_j(1 - u_j)^{r_j - 1}] \, du_1 du_2 \ldots du_j.$$

8.4. (Continuation) Show that, whatever the density f, the scores $q_j^{(u)}$ and $q_j^{(c)}$ satisfy the equations

$$q_j^{(u)} + (r_j - 1)q_j^{(c)} = r_j q_{j-1}^{(c)} \qquad (j = 1, 2, \ldots, d + 1).$$

Hint: Use integration by parts in u_j, together with the relation

$$\frac{d}{du}[(1 - u)\Psi(u)] = -\psi(u).$$

8.5. Weighted log rank statistics, introduced in Section 8.6, have the form

$$T_1(w) = \sum_{j=1}^{d} w_j(d_{1j} - r_{1j}/r_j),$$

where the w_j may be any function of the history of failures and censorings up to τ_j. Rank-regression statistics, introduced in Section

8.7, have the form

$$T_2(q) = \sum_{j=1}^{d} (q_j^{(u)} d_{1j} + q_j^{(c)} m_{1j}),$$

where m_{1j} is the number of members of group 1 in $\mathcal{M}_j$. Show that any test statistic of the form T_2 can be written in the form T_1 by setting

$$w_j = q_j^{(u)} - q_j^{(c)} = r_j(q_{j-1}^{(c)} - q_j^{(c)}) \qquad q_0^{(c)} = 0.$$

Note that these equations also determine the $q_j^{(u)}$ and $q_j^{(c)}$ in terms of the w_j.

8.6. Investigate these relationships for the log rank test and for Prentice's version of the Wilcoxon test. Show by direct calculation that

$$\sum_{j=1}^{d} [q_j^{(u)} + m_j q_j^{(c)}] = 0, \qquad \sum_{j=1}^{d} ([q_j^{(u)}]^2 + m_j[q_j^{(c)}]^2) = \sum_{j=1}^{d} w_j^2(1 - r_j^{-1}).$$

8.7. Use these results and equation (7.15) to evaluate the mean and variance of the statistic T_2 under a permutation model in which the times of failure and censoring are held fixed but in which each of the $n!$ possible labellings of the n subjects is equally likely. Note that this model is applicable only if censoring is independent of the covariate z. Compare the variance with that obtained from the partial likelihood statistic, which does not require this assumption about the censoring.

8.8. Suppose that for each individual there are two alternative ways, denoted by t_1 and t_2, of measuring time. (For example these might in an industrial context be real time and 'usage'.) Each individual follows a trajectory in the north-east quadrant in the (t_1, t_2) plane. A timescale may be called 'complete' if, with time measured on that scale, failure time is independent of the other measure of time. Show how by introducing a new time $t(\theta) = t_1 \cos\theta + t_2 \sin\theta$ and associated explanatory variable $z(\theta) = t_1 \sin\theta - t_2 \cos\theta$, it is possible to find those values of θ, if any, for which the associated timescale is complete and hence in particular to compare the merits of t_1 and t_2.

[Farewell and Cox, 1979]

CHAPTER 9

Several types of failure

9.1 Introduction

In the previous discussion we have not distinguished between different types or modes of failure. We have, however, allowed for the possibility of censoring, withdrawal from study for some reason independent of the failure process, such as preplanned termination of testing. We now consider the different, although related, matter of analysis when there are several types of failure.

We suppose that for each individual there is a failure time T and an indicator random variable V which specifies the type of failure. In the simplest case V takes values 1 and 2 corresponding to just two distinct types of failure. We suppose also that individuals may be censored in the sense outlined above. It is convenient to arrange that each failure is of just one type; this is easily achieved by, if necessary, defining simultaneous failures of type say 1 and 2 as constituting a new type of failure, identified by a separate value of V.

Note particularly that only one failure time is observed for each individual. Multivariate failure time problems, in which two or more failure times are observed on each individual, are deferred to Chapter 10.

One important application is to studies of human mortality. We may be interested in a detailed classification of the causes of death, or we may be content to classify deaths as due or not due to some cause of special interest; it is well known, however, that assignment and interpretation of causes of death can be difficult. In industrial reliability testing, it may be fruitful to classify individuals by the physical mode of failure. Alternatively when the failure of a system is involved, the particular component or subsystem failing may define the type of failure.

A natural generalization is to allow the random variable V to be real-valued or even vector-valued. The general term marked failure is often used. For instance, the mark might correspond to the cost associated with a particular failure; in a medical context the mark might aim to measure the quality of life experienced during a period of treatment.

9.2 Some key relations

(i) Joint distribution

Consider the joint distribution of a random variable (T, V), where T denotes failure time and V, indicating type of failure, takes values $\{1, \ldots, l\}$. The joint distribution can be specified in various equivalent ways. For example, the separate hazard functions

$$h_v(t) = \lim_{\Delta \to 0+} \frac{\mathrm{pr}(t \leqslant T < t + \Delta, V = v | t \leqslant T)}{\Delta} \qquad (9.1)$$

determine, by the addition law of probability, the marginal hazard

$$h(t) = \sum_v h_v(t) \qquad (9.2)$$

and hence the marginal density and survivor functions of T. Given that failure occurs at time t, the conditional probability that the failure is of type v is

$$h_v(t)/h(t),$$

so that the marginal probability that the failure is of type v is

$$\mathrm{pr}(V = v) = \int_0^\infty h_v(t) \exp\left(-\int_0^t h(z)\,dz\right) dt. \qquad (9.3)$$

An important special case has for all v and t

$$h_v(t) = \pi_v h(t), \qquad (9.4)$$

where the π_v's are independent of t and $\sum_v \pi_v = 1$. This is an assumption of proportional hazards, the term being used, however, in a rather different sense from that of Chapters 5–8. Equation (9.4) implies that T and V are independent. A less restrictive assumption requires the proportionality only for $v \in \mathscr{V}^0$, some subset of $\{1, \ldots, l\}$. Thus we might have $h_2(t)/h_1(t)$ independent of t; then, conditionally on failure being of type 1 or 2, T and V are independent.

From a large number of observations on (T, V), the joint distribution, and hence the functions $h_v(t)$, can be estimated arbitrarily closely, no special assumption being required. Conversely any set of nonnegative functions defines a possible distribution for (T, V), with the proviso that if

$$\int_0^\infty h(t)\,dt < \infty$$

the distribution of T is defective.

If V is a continuous random variable or a vector, obvious generalizations of (9.1)–(9.4) hold.

(ii) Competing risks

Quite often interest is focused on one type of failure and we would like to study that type of failure on its own. Possible objectives are to study

(a) the distribution of failure time for, say, type 1 failures, other types of failure having been eliminated;

(b) the comparison of, say, type 1 failures in two or more groups of individuals having different properties for the other types of failure;

(c) the effect on the marginal distribution of failure time of eliminating or reducing the type 1 failures.

The notion of 'eliminating' a type of 'failure' may have a clearcut meaning, as for example when the types of failure refer to separate subsystems. In other cases, as for instance in connection with human survival, very cautious interpretation has to be given.

To approach the objectives (a)–(c), suppose that for each individual there are l notional failure times $T^{(1)}, \ldots, T^{(l)}$; the failure time for type v, all other sources of failure being eliminated, is $T^{(v)}$. Suppose further that the observed random variables (T, V) are given by

$$T = \min(T^{(1)}, \ldots, T^{(l)}), \qquad V = v \text{ if } T = T^{(v)}. \tag{9.5}$$

Such a representation is called one of competing risks. It is assumed that exact equality of two or more of the T can be disregarded, in view of the assumption that only one type of failure occurs.

Now the hazard function (9.1), which is observable, can be written in the present notation

$$h_v(t) = \lim_{\Delta \to 0+} \frac{\mathrm{pr}(t \leqslant T^{(v)} < t + \Delta \,|\, t \leqslant T^{(w)}, w = 1, \ldots, l)}{\Delta}, \tag{9.6}$$

whereas the hazard function $h^{(v)}(t)$ of the random variable $T^{(v)}$ is

$$h^{(v)}(t) = \lim_{\Delta \to 0+} \frac{\mathrm{pr}(t \leqslant T^{(v)} < t + \Delta \,|\, t \leqslant T^{(v)})}{\Delta}. \tag{9.7}$$

The two probabilities in (9.6) and (9.7) are equal for all t and v if the random variables $T^{(1)}, \ldots, T^{(l)}$ are mutually independent. More generally it is enough that the random variables are quasi-independent in that probabilities conditional on

$$t \leqslant T^{(1)}, \ldots, t \leqslant T^{(l)} \tag{9.8}$$

are to be equal to those conditional only on $t \leqslant T^{(v)}$. Full independence would allow unequal $t_1, \ldots, t_l$ in (9.8). It is, however, hard to envisage practical circumstances where quasi-independence holds without full independence, so that the extra generality is of theoretical rather than immediately practical importance. The representation (9.5) with independent (or quasi-independent) components is called the independent competing risk model.

Thus while $h_v(t)$ can always be estimated without special assumptions, $h_v(t)$ can be used directly to determine the hazard, survivor function and density of $T^{(v)}$ only with the appropriate assumption of independence.

When the different $T^{(v)}$ refer to separate subsystems, the assumption of independence may be reasonable. In other instances, for example human survival, the independence is suspect or even implausible. Unfortunately, no direct check of independence is possible with the kind of data under consideration here. For, given an arbitrary distribution for (T, V), the functions $h_v(t)$ are uniquely defined. Under independence, $h_v(t)$ can be used to define an independent competing risk model, with $T^{(v)}$ assigned survivor function

$$\exp\left(- \int_0^t h_v(z)\, dz \right).$$

The model (9.5) then reproduces exactly the specified distribution for (T, V), i.e. there is a unique independent competing risk model corresponding to any given distribution for (T, V).

(iii)   Bounds

Consider the competing risk model (9.5) with, for simplicity, just two types of failure; if we are interested particularly in say $T^{(1)}$, we can always condense all the other types of failure into a single type, defining $T^{(2)}$. Suppose that we want to determine the distribution of $T^{(1)}$ without assuming independence, or any other special condition. By the argument in (ii), the best we can hope for are bounds and these are achieved from the two extreme cases:

(a) for every observed type 2 failure, the unobserved value of $T^{(1)}$ is only slightly greater than the observed failure time;

(b) for every observed type 2 failure, the unobserved value of $T^{(1)}$ is effectively infinite.

This gives two bounds for the survivor function of $T^{(1)}$, the survivor function for all failures and the survivor function of observed type 1 failures, treating failures of other types as corresponding to infinite type 1 failure time.

Of course these bounds will often be too wide to be of value.

(iv) Dependent models

Another approach to the competing risk model (9.5) is to restrict the joint distribution of $T^{(1)}, \ldots, T^{(l)}$, for instance to some simple parametric form. It follows from (9.6) that if $\mathscr{G}(t_1, \ldots, t_l)$ is the survivor function of $T^{(1)}, \ldots, T^{(l)}$, then

$$h_v(t) = -\left[\frac{\partial \log \mathscr{G}(t_1, \ldots, t_l)}{\partial t_v} \right]_{t_1 = \ldots = t_l = t}. \tag{9.9}$$

If $\mathscr{G}(.)$ is defined by a limited number of unknown parameters, then it will often happen that the observable functions $h_v(t)$ determine all the parameters in $\mathscr{G}$, including those that define the dependence between components. A special case (Nádas, 1971) is the multivariate log normal distribution for which both the mean vector and the covariance matrix are determined from the joint distribution of (T, V) and can be estimated from a random sample.

(v) Relation with censoring

Formally we may treat censoring as a type of failure and then the independent competing risk model (9.5) is exactly the model used previously for uninformative censoring. That is, if $T^{(0)}$ is a censoring time and $T^{(1)}$ a failure time, (9.5) becomes

$$T = \min(T^{(0)}, T^{(1)}), \qquad V = \begin{cases} 1 & \text{failure,} \\ 0 & \text{censored.} \end{cases}$$

Thus the statistical procedures developed in the remainder of this chapter are similar to those of earlier chapters. We prefer, however, to treat censoring as distinct from failure, partly to emphasize that censoring need not be random.

9.3 Statistical analysis for independent competing risks

(i) Likelihood function

Suppose that observations are obtained on n independent individuals in the form $(t_1, v_1; z_1), \ldots, (t_n, v_n; z_n)$, where z_i is a vector of explanatory variables for the ith individual. If the individuals form a homogeneous population, then z_i may be suppressed. We take the independent competing risk model, writing $\mathscr{F}_1(t; z), \ldots, \mathscr{F}_l(t; z)$, $h_1(t; z), \ldots, h_l(t; z)$ for the associated survivor and hazard functions. The log likelihood function is

$$\sum_v \sum_i \log \mathscr{F}_v(t_i; z_i) + \sum_v \sum_i \Delta_{iv} \log h_v(t_i; z_i), \qquad (9.10)$$

where $\Delta_{iv} = 1 \, (v = v_i), \Delta_{iv} = 0 \, (v \neq v_i)$. The form (9.10) arises because each individual contributes to the likelihood a factor $\mathscr{F}$ for each type of failure not observed and a factor equal to the density for the type of failure that is observed. Any individuals censored contribute only to the first term.

(ii) Parametric analysis

Suppose that the l underlying distributions are determined by parameter vectors $\theta_1, \ldots, \theta_l$ which vary independently. There is no need for the distributions to be of the same mathematical form or for the dimensions of $\theta_1, \ldots, \theta_l$ to be the same. Then for maximum likelihood and associated methods (9.10) is the sum of l separate terms which can be analysed separately. That for failure type v is

$$\sum_i \log \mathscr{F}_v(t_i; z_i) + \sum_{i:v_i = v} \log h_v(t_i; z_i) \qquad (9.11)$$

and the methods of previous chapters can be used, treating all sources of failure except the vth as censoring. Further, to the first order of asymptotic theory, the parameters $\theta_1, \ldots, \theta_l$ are estimated independently.

The simplest special case of interest arises when the individuals come from a homogeneous population and the underlying survivor functions are exponential with parameters $\rho_1, \ldots, \rho_l$. Then in the absence of censoring, the log likelihood (9.10) is

$$\sum d_v \log \rho_v - t. \sum \rho_v, \qquad (9.12)$$

where d_v is the number of failures of type v and $t.$ is the total time at risk. Thus $\hat{\rho}_v = d_v/t$. and the matrix of second derivatives of the log likelihood is diagonal. More complicated problems are handled similarly.

The main circumstance where the same parameter might enter the specification of more than one type of failure is probably when distributions of similar form are used for several types of failure. For example, Weibull distributions might be used and the possibility considered that the distributions all have the same index.

(iii)   Nonparametric analysis

For nonparametric and semi-nonparametric models, the results of previous chapters carry over directly so long as the component random variables in (9.5) are independent or quasi-independent, and any parameters are separate for the different types of failure.

For example, product-limit estimates can be formed separately for the different types of failure and their statistical fluctuations treated as asymptotically independent. Note that the resulting statistics in effect estimate $h_v(t)$ of (9.1) and the survivor function based on this, without any special assumption of independence or quasi-independence of components. These last assumptions are needed to give $h_v(t)$ an interpretation very specific to a single failure type. Also for the study of dependence on explanatory variables, via the proportional hazards model of Chapter 5, separate partial likelihood functions can be constructed for each type of failure and, so long as each type of failure has a separate vector of regression coefficients, the resulting vectors of estimates are asymptotically independent.

9.4   Proportional hazards

The emphasis in Section 9.3 has been on the study of failure types one at a time. Sometimes it may be required to examine the relation between hazard functions referring to different types of failure.

This can be tackled in various ways, for example by plotting estimated survivor functions or log survivor functions, or by smoothing and plotting hazards or log hazards, or via some parametric formulation. Here we consider the comparison of just two types of failure through the representation

$$h_2(t) = e^{\beta_0 + \beta_1 t} h_1(t). \tag{9.13}$$

When $\beta_1 = 0$, we have the proportional hazards form (9.4).

Given independent observations $(t_1, v_1), \ldots, (t_n, v_n)$ on uncensored individuals, we may condition on $t_1, \ldots, t_n$ and use the consequence of (9.13) that

$$\frac{\text{pr}(V_i = 2 | T_i = t_i)}{\text{pr}(V_i = 1 | T_i = t_i)} = e^{\beta_0 + \beta_1 t_i}. \tag{9.14}$$

Thus methods for binary logistic regression (Cox, 1970) are directly applicable. In particular the null hypothesis of proportional hazards is tested by regarding

$$\sum_{i: v_i = 1} t_i \tag{9.15}$$

as the total of a random sample of size d_1 drawn without replacement from the finite population $\{t_1, \ldots, t_n\}$; here d_1 is the number of individuals experiencing type 1 failure. The mean and variance of (9.15) under the null hypothesis are respectively

$$d_1 \bar{t}, \qquad d_1 (n - d_1) \sum (t_i - \bar{t})^2 / [n(n-1)], \tag{9.16}$$

where $\bar{t} = \sum t_i / n$.

Explanatory variables can be included in (9.13) and (9.14) in the

Table 9.1 *Frequency distributions of times of failure of radio transmitters*

Time (h)	Type I (confirmed)	Type II (unconfirmed)	Total
0–	26	15	41
50–	29	15	44
100–	28	22	50
150–	35	13	48
200–	17	11	28
250–	21	8	29
300–	11	7	18
350–	11	5	16
400–	12	3	15
450–	7	4	11
500–	6	1	7
550–	9	2	11
600–629	6	1	7
Not failed at 630 h	—	—	44
Total	218	107	369

usual way. The conditioning on $t_1, \ldots, t_n$ is appropriate when the function $h_1(t)$ is regarded as unknown. If $h_1(t)$ can reasonably be specified parametrically, then direct use of maximum likelihood will be possible.

As an example, consider the data of Mendenhall and Hader (1958) given in grouped form in Table 9.1. These concern failures of radio transmitters, there being two types of failure. Fig. 9.1 shows the proportion of failures of type 1 plotted against the centre of the age group. To test the constancy of the hazard ratio $h_2(t)/h_1(t)$ from the grouped data we may initially examine the chi-squared statistic from the 2×13 contingency table of frequencies of failures, type $\times$ age group, on which Fig. 9.1 is based. This gives 9.37 with 12 degrees of freedom, giving no suggestion of undue variation in the proportion of failures of, say, type 1, $h_1(t)/[h_1(t) + h_2(t)]$. Such a test is, of course, insensitive to the kind of smooth changes of most interest. For the test based on (9.15) and (9.16) we find the total, or equivalently the mean, of the type 1 failure times; the mean is 229.7, the null hypothesis mean and standard deviation being 218.5 and 6.18 respectively, so that the discrepancy is 1.8 standard errors. There is thus some evidence, by no

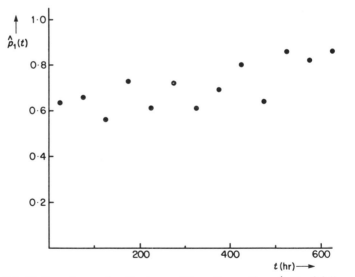

Fig. 9.1. Failure-times of radio transmitters. Proportion, $\hat{p}_1(t)$ of failures of type 1 versus time, t.

means conclusive, that $h_1(t)/[h_1(t) + h_2(t)]$ increases with time. Fig. 9.1 shows that the evidence for this increase depends quite strongly on the values above 500 h.

9.5 Marked failures

We now consider briefly some ideas useful when, associated with each failure, there is a further measurement, additional to the failure time, this additional measurement being in some sense a 'response', to be interpreted in conjunction with the failure time. We call such a further measurement a mark, in line with the terminology used in the theory of point processes. A simple example of a mark, already studied in Section 9.2, specifies the type of failure. Other examples include, in an industrial context, a measure of the work performed by a component or system during its lifetime or of the cost associated with its failure. In some contexts the mark is defined for all individuals, but often censored individuals will have no observed mark. In general, we consider the mark as a random variable V, in general a vector. It will usually be sensible to begin by examining the marginal distribution of V, and its possible dependence on explanatory variables, by whatever formal or informal techniques seem suitable in the light of the nature of V, e.g. binary, discrete, continuous, approximately normal, and so on. Here, however, we concentrate on the interrelationship between V and failure time, T, in situations where explanatory variables are available.

Three extreme idealized cases can be distinguished:

(i) given the explanatory variables z, V and T are independent;
(ii) V 'drives' the process in the sense that the conditional distribution of T given $V = v$ does not depend on z;
(iii) T 'drives' the process in the sense that the conditional distribution of V given $T = t$ does not depend on z.

Possibilities (ii) and (iii) can be examined by formal tests of independence, i.e. by testing for dependence on z for case (ii) in a model including both v and z as explanatory variables influencing T, and in case (iii) in a model for V with both t and z as explanatory variables. In the latter case, the handling of censored failure times needs care. If the mark is not observed on the censored individuals, there is nothing that can be done with censored individuals. If a mark is observed on a censored individual and an explicit model is constructed for the joint distribution of (T, V), then a factor for the likelihood can be computed: one reasonably simple such model is to have, possibly after transformation, (T, V) having a bivariate log normal distribution incorporating linear regression on the explanatory variables.

A further possibility, related to but different from a marked process, is the use of a surrogate variable on censored individuals as in effect a predictor of failure time. For example, in a clinical trial in which censored individuals are those who have not experienced a 'critical event' by the end of the trial, there might be available some observation or set of observations that, after suitable calibration, could be used to predict failure time and which thereby provide some information about the effect of the explanatory variables. To be useful, the distribution of the surrogate variables must depend in a known way on the regression parameter β; see Exercise 9.7.

In some cases the mark is, at least in principle, measured continuously over the lifetime of the individual; we write $V(s)$ for the value of the mark at time s. Rate of output achieved in an industrial context or quality of life in a medical context are examples. In some analyses such variables would be considered as time-dependent covariates, but here we regard them as responses.

One important difference between the two examples is that in the first, but not in the second, the integral of $V(.)$ over the life of the individual has a clear physical interpretation. Thus an analysis of the effect of explanatory variables on the mean or integral of $V(.)$ and its relation to failure time is a natural starting point for any analysis. The time-series structure (trend, etc.) of $V(.)$ could, of course, also be studied. Note that individuals censored with respect to time would have some information available on $V(.)$.

In the second example, however, the integral of $V(.)$ no longer has a very clear meaning and, in the presence of appreciable variability, even the mean has to be interpreted cautiously. For example, with quality measured on some numerical scale, it will be rash to assume that 2 units of time at quality level 4 is better (or worse) than 5 units of time spent at quality level 2. The safest procedure is probably to set a number of quality levels and to obtain for each individual the number of units of time spent at or above each quality level. Now the total 'occupation time' defined in this way can be regarded as a 'failure time' measured on a new scale. Therefore for a sample of individuals with the same values of the explanatory variables, the product-limit estimates of the distribution functions at various levels of quality provide at least a reasonable first summary of censored data.

Bibliographic notes, 9

The ideas in this chapter have a long history, going back to Daniel Bernouili's (1760) study of the potential consequences of smallpox inoculation. For a historical review of competing risk theory, see Seal (1977), and for an account of the extensive actuarial work on multiple types of failure, Elandt-Johnson and Johnson (1980, Part 3).

The nontestability of dependence in the competing risk model was noted by Cox (1959, 1962) and studied in detail by Tsiatis (1975). Aalen (1976) has emphasized the direct interpretation of the estimable functions $h_v(t)$. The explicit requirement of quasi-independence in the competing risk model was noted in the Melbourne thesis of R. Fisher. See also Williams and Lagakos (1977) and Langberg et al. (1981). For a review of competing risk theory, see Gail (1975).

Bounds on the underlying distribution function were explicitly formulated by Peterson (1977).

Models with explicitly dependent underlying failure times were studied by Nádas (1970, 1971), David and Moeschberger (1978), Moeschberger (1974) and Lagakos and Williams (1978).

Formal examination of proportionality of hazards of different types was studied by Cox (1959).

General marked failures have not been much studied; for work on marked point processes, see Cox and Isham (1980, Section 5.5).

Further results and exercises, 9

9.1. Develop the asymptotic relative efficiency of the test of proportional hazards of Section 9.4 when the underlying distributions are of the Weibull form, so that proportionality of hazards is equivalent to equality of the Weibull indices.

9.2. Extend the test of Section 9.4 to the testing of the proportionality of l hazard functions, $l > 2$.

9.3. Discuss carefully how it is that formally the same procedure that arises in Section 9.4 can be relevant when

(i)   both 'time' and 'type' are random;

(ii)   the 'times' are fixed and 'type' is a random binary response;
(iii) samples of preassigned size are drawn from two distributions.

Is there a physical interpretation of the formulation in which both 'times' and the total numbers of the two types are to be regarded as fixed *a priori*? Comment on the corresponding problems arising in the analysis of 2×2 contingency tables.

9.4. Suppose that on failure a univariate continuously distributed mark is observed. Assuming no censoring, suggest how a large body of data could be analysed to develop simple representations in which

(a) failure time and mark are treated on a symmetrical footing;
(b) failure time is dependent on mark;
(c) mark is dependent on failure time.

How would these procedures be affected by censoring, the mark being unobserved on censored individuals? How could explanatory variables be incorporated?

9.5. Suppose that failure time T is exponentially distributed with parameter ρ and that, conditionally on $T = t$, the mark V is normally distributed with mean $\xi + \eta t$ and variance σ^2. Prove that the marginal density of V is

$$\lambda^{-1} \Phi \left(\frac{v - \xi}{\sigma} - \frac{\sigma}{\lambda} \right) \exp \left(\frac{\sigma^2}{2\lambda^2} - \frac{v - \xi}{\lambda} \right),$$

where $\lambda = \eta/\rho$ and that its mean and variance are $\xi + \lambda$ and $\sigma^2 + \lambda^2$.

Show how the parameters $(\rho, \xi, \eta, \sigma^2)$ can be estimated by standard methods from complete and from censored data, at least when V is missing for censored individuals. How would the model be analysed if V is observed for censored individuals? Suggest how the model can be generalized to represent the effect of explanatory variables.

9.6. Repeat the discussion of Exercise 9.5 taking V to be binary with conditional probability of 'success' $\Phi(\xi + \eta t)$.

9.7. Suppose that for the ith individual, failure time is exponentially distributed with rate parameter ρ_i, where $\log \rho_i = \beta^T z_i$, and where z_i is a $p \times 1$ vector of fixed explanatory variables. Thus, by Section 6.2(i), the information matrix in the absence of censoring is $z^T z$, where for n individuals z is the $n \times p$ matrix of explanatory variables. Suppose that the potential censoring time for the ith individual is exponentially distributed with parameter κ_i, so that the probability of censoring is $\pi_i = \kappa_i/(\kappa_i + \rho_i)$. Prove that the information matrix for β is now $z^T(I^* - \pi)z$, where $\pi = \text{diag}(\pi_1, \ldots, \pi_n)$ and I^* is the identity matrix. Suppose now further that for each censored individual a supplementary random variable is available of the form $V_i = \tilde{V}_i/W_i$, where $\tilde{V}_i$ is the unobserved remaining lifetime of the individual and W_i is an independent error component having a gamma distribution. Compute the distribution of V_i and thence the information matrix for β and the parameters of the gamma distribution. Find the proportion of the lost information about β that is recovered by the use of V_i, showing in particular that if W_i is exponentially distributed the proportion is about $1/3$.

9.8. To what extent are the assumptions about the quantity V_i of Exercise 9.7 capable of empirical test? Suppose that for all individuals a prognostic index is found recorded at fairly frequent intervals throughout the individual's progress. Discuss how the prognostic index for the individuals who in fact fail might be used to calibrate the values for the censored individuals, thereby overcoming some of the obvious objections to the procedure of Exercise 9.7.

[Cox, 1983]

CHAPTER 10

Bivariate survivor functions

10.1 Introduction

The identifiability problems in competing risk data arise because only the first failure to occur on any subject is observed, this effectively censoring the remaining time or times to failure. In industrial reliability this corresponds to units arranged in series, so that the entire system fails as soon as any unit fails. If two units are arranged in parallel, then the entire system can function so long as either unit functions. The two times to failure, $T^{(1)}$ and $T^{(2)}$, can both be observed and all aspects of their joint distribution studied. Interest may centre on whether a failure of the first type at $t^{(1)}$ increases the hazard in $t \geqslant t^{(1)}$ for failures of the second type. This question can be answered by the methods of Chapter 8, regarding the failures of the first type as defining time-dependent covariates which possibly influence the hazard for failures of the second type.

The present chapter is mainly concerned with applications which do not involve the singling out of one variable as a response, that is with studies of correlation rather than of regression. Then the full joint distribution of $T^{(1)}$ and $T^{(2)}$ becomes of interest. Consideration of the process unfolding in time directs attention to the hazard functions

$$h^{(i)}(t) = \lim_{\Delta \to 0} \frac{1}{\Delta} \operatorname{pr}(T^{(i)} < t + \Delta \,|\, T^{(1)} \geqslant t, T^{(2)} \geqslant t)$$

of failures of type i at t given no failure before t and, for $i = 1, 2$, $i + j = 3$ and $t' \geqslant t$,

$$\bar{h}^{(i)}(t' \,|\, t) = \lim_{\Delta \to 0} \frac{1}{\Delta} \operatorname{pr}(T^{(i)} < t' + \Delta \,|\, T^{(i)} \geqslant t', T^{(j)} = t)$$

of a failure of type i at t' given a previous failure of the other type j at t. These four functions determine the joint distribution of $(T^{(1)}, T^{(2)})$ if this is continuous. The joint density is for $t^{(1)} < t^{(2)}$

$$f_{T^{(1)},T^{(2)}}(t^{(1)}, t^{(2)}) = \exp\left(-\int_0^{t^{(1)}} [h^{(1)}(t) + h^{(2)}(t)]\,dt - \int_{t^{(1)}}^{t^{(2)}} \bar{h}^{(2)}(t|t^{(1)})\,dt \right)$$
$$\times\; h^{(1)}(t^{(1)})\bar{h}^{(2)}(t^{(2)}|t^{(1)}) \tag{10.1}$$

with a similar expression if $t^{(1)} > t^{(2)}$.
The possibility of simultaneous occurrence of the two types of failure, $T^{(1)} = T^{(2)}$, can be handled by the inclusion of a delta function component in $\bar{h}$.

10.2 A 'shock' model

One approach to the construction of bivariate survival distributions is to generalize the lack of memory property of the exponential distribution. A natural bivariate analogue of this property is that for $s, t^{(1)}, t^{(2)} > 0$

$$\mathrm{pr}(T^{(1)} \geqslant s + t^{(1)}, T^{(2)} \geqslant s + t^{(2)} \,|\, T^{(1)} \geqslant s, T^{(2)} \geqslant s)$$
$$= \mathrm{pr}(T^{(1)} \geqslant t^{(1)}, T^{(2)} \geqslant t^{(2)}), \tag{10.2}$$

so that the joint density of the residual lifetimes $T^{(1)} - s$, $T^{(2)} - s$, given that no failure occurs before s, does not depend on s. In terms of the joint survivor function

$$\mathscr{F}(t^{(1)}, t^{(2)}) = \mathrm{pr}(T^{(1)} \geqslant t^{(1)}, T^{(2)} \geqslant t^{(2)}),$$

(10.2) becomes

$$\mathscr{F}(s + t^{(1)}, s + t^{(2)}) = \mathscr{F}(s, s)\mathscr{F}(t^{(1)}, t^{(2)}). \tag{10.3}$$

By applying the factorization (10.3) with $t^{(1)} = t^{(2)}$, we deduce that the function $A(s) = \mathscr{F}(s, s)$ satisfies the functional equation

$$A(s + u) = A(s)A(u),$$

from which it follows that, for some ρ, $A(s) = e^{-\rho s}$. This result can equivalently be deduced by noting that $A(s)$ is the survivor function of $\min(T^{(1)}, T^{(2)})$, which by (10.2) must have the univariate lack of memory property.

If, in addition to (10.2), it is required that the marginal densities of $T^{(1)}$ and $T^{(2)}$ each be exponential, say with parameters $\rho^{(1)}, \rho^{(2)}$, further characterization of the joint distribution is possible. We have immediately

$$\mathcal{F}(t^{(1)}, t^{(2)}) = \begin{cases} \exp[-\rho t^{(1)} - \rho^{(2)}(t^{(2)} - t^{(1)})] & (t^{(1)} < t^{(2)}), \\ \exp[-\rho t^{(2)} - \rho^{(1)}(t^{(1)} - t^{(2)})] & (t^{(1)} \geqslant t^{(2)}). \end{cases} \qquad (10.4)$$

In terms of new parameters $\rho_1 = \rho - \rho^{(2)}$, $\rho_2 = \rho - \rho^{(1)}$ and $\rho_{12} = \rho^{(1)} + \rho^{(2)} - \rho$, (10.4) can be re-expressed as

$$\mathcal{F}(t^{(1)}, t^{(2)}) = \exp[-\rho_1 t^{(1)} - \rho_2 t^{(2)} - \rho_{12} \max(t^{(1)}, t^{(2)})]. \qquad (10.5)$$

For (10.5) to represent a proper joint survivor function, $\rho_1 \geqslant 0$, $\rho_2 \geqslant 0$ and $\rho_{12} \geqslant 0$ (Exercise 10.2). It then corresponds to a system in which three types of shock can occur, resulting respectively in failures of the first, of the second and of both types simultaneously. The times to each type of shock are independently exponentially distributed with parameters ρ_1, ρ_2 and ρ_{12} respectively.

10.3 A continuous bivariate survival model

The shock model discussed in the previous section leads to a bivariate distribution which is not absolutely continuous. There is a singularity along the line $t^{(1)} = t^{(2)}$ in the $(t^{(1)}, t^{(2)})$ plane. In some applications, where the two times to failure are related only indirectly and may even be measured in different units, such discontinuities will be inappropriate. For example, $t^{(1)}$ may be the time to discovery of a defect in a machine and $t^{(2)}$ the subsequent time to its failure. In familial studies of disease incidence, $t^{(1)}$ may be the age at death from a specified disease of a father, $t^{(2)}$ the age at death from the same disease of his son, other causes of death in both cases acting as censoring.

The distinction between this and the preceding section corresponds to that between multidimensional point processes, where each pair of times is represented by a single point in a plane, and multivariate processes, where the representation would consist of two points of different types on a single time axis. A number of desirable properties for families of two-dimensional continuous survivor functions, as they might be called, can be listed.

(i) The association between $T^{(1)}$ and $T^{(2)}$ is governed by a single parameter ϕ which has a simple physical interpretation.

(ii) The marginal survivor functions $\mathcal{F}_1(t^{(1)}) = \mathcal{F}(t^{(1)}, 0)$ and $\mathcal{F}_2(t^{(2)}) = \mathcal{F}(0, t^{(2)})$ can be specified arbitrarily and, if desired, parameterized separately from ϕ.

(iii) Either negative or positive association should be permissible, and the special cases of independence and the Fréchet bounds

$$\mathscr{F}_U(t^{(1)}, t^{(2)}) = \min[\mathscr{F}_1(t^{(1)}), \mathscr{F}_2(t^{(2)})], \qquad (10.6)$$

$$\mathscr{F}_L(t^{(1)}, t^{(2)}) = \max[0, \mathscr{F}_1(t^{(1)}) + \mathscr{F}_2(t^{(2)}) - 1] \qquad (10.7)$$

are achievable within the family. These bounds give the joint survivor functions corresponding respectively to the greatest possible positive and negative association between $T^{(1)}$ and $T^{(2)}$ consistent with the specified marginal distributions. The two joint distributions are singular, placing all their mass along the respective curves $\mathscr{F}_1(t^{(1)}) = \mathscr{F}_2(t^{(2)})$ and $\mathscr{F}_1(t^{(1)}) + \mathscr{F}_2(t^{(2)}) = 1$ in the $(t^{(1)}, t^{(2)})$ plane.

(iv) Reasonably simple parametric and semiparametric procedures are available for estimating ϕ, even in the presence of right censoring in either or both components.

A family which satisfies most of these requirements, except that it cannot represent negative association, was proposed by Clayton (1978). He postulated a relation of the form

$$h_{T^{(1)}}(t^{(1)} | T^{(2)} = t^{(2)}) = (1 + \phi) h_{T^{(1)}}(t^{(1)} | T^{(2)} \geq t^{(2)}) \qquad (10.8)$$

between the hazard functions of the conditional distribution of $T^{(1)}$ given $T^{(2)} = t^{(2)}$ and given $T^{(2)} \geq t^{(2)}$. These hazard functions differ from those introduced in Section 10.1. In particular, there is no discontinuity in the definition along the line $t^{(1)} = t^{(2)}$. It is easily seen that (10.8) is equivalent to a similar condition with the roles of $T^{(1)}$ and $T^{(2)}$ interchanged.

For any continuous bivariate survivor function $\mathscr{F}(t^{(1)}, t^{(2)})$, with marginals $\mathscr{F}_1(t^{(1)}) = \mathscr{F}(t^{(1)}, 0)$ and $\mathscr{F}_2(t^{(2)}) = \mathscr{F}(0, t^{(2)})$, the conditional survivor function of $T^{(1)}$ given $T^{(2)} \geq t^{(2)}$ is

$$\mathscr{F}(t^{(1)}, t^{(2)}) / \mathscr{F}_2(t^{(2)}),$$

and that of $T^{(1)}$ given $T^{(2)} = t^{(2)}$ is

$$\frac{\partial \mathscr{F}(t^{(1)}, t^{(2)})}{\partial t^{(2)}} \bigg/ \frac{\partial \mathscr{F}_2(t^{(2)})}{\partial t^{(2)}}.$$

Thus in view of (2.7) we have that

$$h_{T^{(1)}}(t^{(1)} | T^{(2)} = t^{(2)}) = -\frac{\partial}{\partial t^{(1)}} \log\left[-\frac{\partial \mathscr{F}(t^{(1)}, t^{(2)})}{\partial t^{(2)}}\right], \qquad (10.9)$$

$$h_{T^{(1)}}(t^{(1)} \mid T^{(2)} \geq t^{(2)}) = -\frac{\partial}{\partial t^{(1)}} \log \mathcal{F}(t^{(1)}, t^{(2)}). \qquad (10.10)$$

Substitution of (10.9) and (10.10) into (10.8) and integration over $(0, t_1)$ yields

$$\log\left[-\frac{\partial \mathcal{F}(t^{(1)}, t^{(2)})}{\partial t^{(2)}} \right] - \log\left[-\frac{\partial \mathcal{F}_2(t^{(2)})}{\partial t^{(2)}} \right]$$
$$= (1 + \phi)[\log \mathcal{F}(t^{(1)}, t^{(2)}) - \log \mathcal{F}_2(t^{(2)})].$$

Exponentiation gives

$$\frac{\dfrac{\partial \mathcal{F}(t^{(1)}, t^{(2)})}{\partial t^{(2)}}}{[\mathcal{F}(t^{(1)}, t^{(2)})]^{1+\phi}} = \frac{\dfrac{\partial \mathcal{F}_2(t^{(2)})}{\partial t^{(2)}}}{[\mathcal{F}_2(t^{(2)})]^{1+\phi}} \qquad (10.11)$$

and a second integration, over $(0, t^{(2)})$, gives

$$[1/\mathcal{F}(t^{(1)}, t^{(2)})]^\phi - [1/\mathcal{F}_1(t^{(1)})]^\phi = [1/\mathcal{F}_2(t^{(2)})]^\phi - 1.$$

so that the bivariate survivor function $\mathcal{F}$ takes the form

$$\mathcal{F}(t^{(1)}, t^{(2)}) = ([1/\mathcal{F}_1(t^{(1)})]^\phi + [1/\mathcal{F}_2(t^{(2)})]^\phi - 1)^{-1/\phi}. \qquad (10.12)$$

As $\phi \to 0$ this gives $\mathcal{F}(t^{(1)}, t^{(2)}) \to \mathcal{F}_1(t^{(1)}) \mathcal{F}_2(t^{(2)})$, corresponding to independence between $T^{(1)}$ and $T^{(2)}$, and as $\phi \to \infty$, the upper Fréchet bound (10.6) is obtained.

The family has a 'random effects' interpretation, in which the association between $T^{(1)}$ and $T^{(2)}$ is explained by their common dependence on an unobserved variable W. Specifically, suppose that W has a gamma density

$$f_W(w) \propto w^{\phi^{-1}-1} e^{-w}, \qquad (10.13)$$

and suppose that, conditionally on $W = w$, $T^{(1)}$ and $T^{(2)}$ are independent, with survivor functions

$$\mathcal{F}_j^*(t \mid w) = \exp(w\{1 - [\mathcal{F}_j(t)]^{-\phi}\}) \qquad (j = 1, 2), \qquad (10.14)$$

so that $T^{(1)}$ and $T^{(2)}$ each satisfy a proportional hazards model with multiplier w. Then the unconditional joint survivor function of $T^{(1)}$ and $T^{(2)}$ is

$$\int_0^\infty \mathcal{F}_1^*(t^{(1)} \mid w) \mathcal{F}_2^*(t^{(2)} \mid w) f_W(w)\, dw = \mathcal{F}(t^{(1)}, t^{(2)}), \qquad (10.15)$$

the function (10.12).

Maximum likelihood estimation of ϕ when the marginal survivor functions are parameterized, say as exponential distributions with parameters ρ_1 and ρ_2, is in principle straightforward, although there may be difficulties near the boundary $\phi = 0$ of the parameter space. Goodness of fit could be assessed by fitting a more general family in which ϕ may depend on $t^{(1)}$ and $t^{(2)}$.

10.4 Testing for independence

Tests for dependence between two variables when either or both may be subject to censoring have been considered by many authors. Tests based on parametric or semiparametric models such as those discussed in the previous sections have obvious advantages in terms of interpretability and the possibility of estimating the degree of dependence (in some meaningful sense) should the null hypothesis be rejected.

Alternatively, tests can be based on the null distribution of a descriptive measure of association between the two components. An obvious candidate when censoring is present is Kendall's coefficient of concordance, τ. For an uncensored sample $(T_i^{(1)}, T_i^{(2)}; i = 1, 2, \ldots, n)$, τ is the proportion of concordant pairs (i, j), that is pairs for which

$$(T_i^{(1)} - T_j^{(1)})(T_i^{(2)} - T_j^{(2)}) > 0$$

minus the proportion of discordant pairs, that is pairs where the reverse inequality holds. This can be modified for censored data, rather as Gehan modified the Wilcoxon test, by counting only definite concordances and definite discordances. More powerful tests can be obtained by the inclusion of partial information from those pairs (i, j) which cannot be classified as a definite concordance or definite discordance.

If the censoring mechanisms for the two components are independent, or if there is censoring in only one component, an exact permutation distribution of the test statistic under the null hypothesis can be obtained. Otherwise, recourse must be had to asymptotic theory. In some circumstances, the first component $T^{(1)}$ of the bivariate failure time may determine the potential censoring time $C^{(2)}$ for the second component. This would apply to the machine example considered in Section 10.3 if the total observation time were fixed.

A more systematic approach to the derivation of tests for association is given by consideration of the bivariate accelerated life model, with

$$\log T^{(2)} = \beta_2 X + \varepsilon^{(2)}, \quad \log T^{(1)} = \beta_1 X + \varepsilon^{(1)}, \qquad (10.16)$$

where the random variables X, $\varepsilon^{(1)}$ and $\varepsilon^{(2)}$ are mutually independent. In this model, the association between $T^{(1)}$ and $T^{(2)}$ is explained by their common dependence on X. Locally most powerful rank tests against specific alternative hypotheses can be derived from (10.16).

10.5 Nonparametric estimation

Fully nonparametric estimation of a bivariate survivor function when either or both components may be censored is best viewed as an application of the EM algorithm, which will be discussed in the next chapter. In the most general case, there are four possible types of outcome that may be observed, namely double failures ($T^{(1)} = t^{(1)}$, $T^{(2)} = t^{(2)}$) corresponding to points in the $(t^{(1)}, t^{(2)})$ plane, single failures ($T^{(1)} = t^{(1)}, T^{(2)} > c^{(2)}$) and ($T^{(1)} > c^{(1)}, T^{(2)} = t^{(2)}$) corresponding to halflines parallel to the $t^{(2)}$ and $t^{(1)}$ axes respectively, and double censorings ($T^{(1)} > c^{(1)}$, $T^{(2)} > c^{(2)}$) corresponding to quadrants. The notational aspects of the problem become quite complex and the estimator must be found by iteration.

Simplification is possible when there is censoring in only one component, or if the two components in any pair always have the same potential censoring time. The latter would normally hold for processes which unfold in time in the manner discussed in Section 10.1. The distribution of $T = \min(T^{(1)}, T^{(2)})$ can then be estimated by the usual product-limit method (Chapter 4). For each observed value $t^{(1)}$ of T for which $T^{(2)} \geqslant T^{(1)}$, the conditional distribution of $T^{(2)}$ given $T^{(1)} = t^{(1)}$, $T^{(2)} \geqslant T^{(1)}$ may be estimated the same way, and similarly for values $t^{(2)}$ of T for which $T^{(1)} > T^{(2)}$. The three estimators may be combined to give a single noniterative estimator of the joint distribution.

10.6 Tests for equality of marginal distribution

In the previous sections the association between $T^{(1)}$ and $T^{(2)}$ has been of interest in itself. Often, however, this association arises because of pair-matching in a study where the main objective is the investigation of the relationship between the marginal distributions of $T^{(1)}$ and $T^{(2)}$. To formulate an appropriate generalization of the proportional hazards model, suppose that for the ith pair the hazard function for the first member is, for some constant θ, $h_{1i}(t) = \theta h_i(t)$, and for the second member $h_{2i}(t) = h_i(t)$, where $h_i(t)$ is an underlying baseline hazard function assumed unknown and different for every pair. A conditional likelihood is obtained by considering only those pairs for which a failure is observed while both members are still at risk.

If, in fact, the functions $h_i(t)$ can be specified, say up to proportionality factors ϕ_i, as $h_i(t) = \phi_i h_0(t)$ for some common function $h_0(t)$, appreciable recovery of 'interblock' information may be possible.

Bibliographic notes, 10

The expression (10.1) of the joint distribution in terms of the hazard functions $h^{(i)}$ and $\bar{h}^{(i)}$ was given by Cox (1972). The bivariate shock model of Section 10.2 is due to Marshall and Olkin (1967). Clayton (1978) introduced the model described in Section 10.3 in connection with familial studies of disease incidence. Parametric and semiparametric estimation within this model is discussed by Clayton (1978) and Oakes (1982a), although a fully satisfactory nonparametric procedure has not yet been found. Tests for independence based on Kendall's (1938) coefficient of concordance are discussed by Brown et al. (1974), Weier and Basu (1980) and Oakes (1982b). The permutation theory is based on the results of Daniels (1944). Cuzick (1982) investigated bivariate rank tests based on (10.16). Holt and Prentice (1974) gave the conditional likelihood for the matched pairs model of Section 10.6. See Wild (1983) for further discussion.

Further results and exercises, 10

10.1. Show that if (10.2) is strengthened to

$$\text{pr}(T^{(1)} \geqslant s^{(1)} + t^{(1)}, T^{(2)} \geqslant s^{(2)} + t^{(2)} \mid T^{(1)} \geqslant s^{(1)}, T^{(2)} \geqslant s^{(2)})$$
$$= \text{pr}(T^{(1)} \geqslant t^{(1)}, T^{(2)} \geqslant t^{(2)}) \qquad (s^{(1)}, s^{(2)}, t^{(1)}, t^{(2)} > 0),$$

then, without any further condition, it follows that $T^{(1)}$ and $T^{(2)}$ are independent and exponentially distributed.

10.2. Show that (10.5) can represent a bivariate survivor function if and only if $\rho_1 \geqslant 0$, $\rho_2 \geqslant 0$, $\rho_{12} \geqslant 0$ with strict inequality either for ρ_{12} or both ρ_1 and ρ_2.

10.3. Investigate the parameterization of (10.12) in which the marginal survivor functions $\mathscr{F}_1$ and $\mathscr{F}_2$ are exponential, with parameters $\rho^{(1)}$ and $\rho^{(2)}$. Show that the expected information about ϕ from the likelihood in $\phi, \rho^{(1)}$ and $\rho^{(2)}$ may be evaluated in terms of the trigamma function

$$\psi'(x) = \sum_{n=0}^{\infty} \frac{1}{(n+x)^2}.$$

[Oakes, 1982b]

10.4. Suppose that $T^{(1)}$ and $T^{(2)}$ have an absolutely continuous joint distribution, and that the conditional distributions of $T^{(2)}$ given $T^{(1)}$ and of $T^{(1)}$ given $T^{(2)}$ are both of the proportional hazards form, i.e.

$$h_{T^{(2)}}(t^{(2)} \mid T^{(1)} = t^{(1)}) = a_{11}(t^{(2)}) a_{12}(t^{(1)}),$$
$$h_{T^{(1)}}(t^{(1)} \mid T^{(2)} = t^{(2)}) = a_{21}(t^{(2)}) a_{22}(t^{(1)}),$$

for some functions $a_{ij}(t) \geqslant 0$ $(i, j = 1, 2)$.

Show that, without loss of generality, these functions may be taken to satisfy

$$a_{11}(t) = a'_{21}(t), \qquad a_{22}(t) = a'_{12}(t), \qquad a_{21}(0) = a_{12}(0) = \eta > 0,$$

and determine the joint density in terms of $a_{11}(t)$, $a_{22}(t)$ and η.

10.5. Use the random effects interpretation given in (10.13) and (10.14) to extend the model (10.12) to multivariate survivor functions i.e. with more than two components. Comment on the generality or lack of it, of this model.

CHAPTER 11

Self-consistency and the EM algorithm

11.1 Introduction

As mentioned in Chapter 1, censoring can be viewed as an example of incomplete or missing data. The failure time T_i of a censored subject is not observed, and is known only to exceed the censoring time c_i. Many of the techniques described in this book, especially those that are derived from or can be interpreted as the maximization of a likelihood function, are specializations of a very general concept, called by some authors the missing information principle, and by others, self-consistency. More recently, the term expectation–maximization (EM) algorithm has become popular. This name is especially appropriate to applications involving exponential families, as an iterative procedure for maximizing the log likelihood is obtained. The principle is most useful when the log likelihood

$$l_0(\phi) = l_0(\phi; T) \tag{11.1}$$

of the data $T = (T_1, \ldots, T_n)$ that would be observed if there were no censoring, has an appreciably simpler functional form than the log likelihood

$$l(\phi) = l(\phi; x, v) \tag{11.2}$$

of the data (x, v) that are actually observed. We keep the notation T and (x, v) from Chapter 1 to emphasize the particular application to survival analysis, but the principle applies also to arbitrarily grouped or missing observations and in many other contexts. There is no requirement that the survival times T_i have identical distributions, so that in particular l_0 and l may depend on explanatory variables z, as described in Chapters 5–8. It is important that concavity of l_0 in ϕ does not in general imply concavity of l.

The EM algorithm hinges on the function $Q(\phi', \phi)$ given by

$$Q(\phi', \phi) = E[l_0(\phi'; T)|x, v; \phi],\qquad (11.3)$$

the conditional expectation of the log likelihood based on T, given the observations (x, v). It is essential to distinguish the two arguments of Q. Thus ϕ' is an argument of the full log likelihood l_0, whereas ϕ is the parameter of the conditional distribution of T given (x, v), which is used to calculate the conditional expectation. In terms of the function Q, the two steps of the EM algorithm may be defined as follows:

> *Expectation step*: Given the current estimate $\hat{\phi}_j$ of ϕ, calculate $Q(\phi', \hat{\phi}_j)$, as a function of the dummy argument ϕ'.
> *Maximization step*: Determine a new estimate $\hat{\phi}_{j+1}$ as the value of ϕ' which maximizes $Q(\phi', \hat{\phi}_j)$.

Under fairly general conditions the sequence $\{\hat{\phi}_j\}$ will converge to the value $\hat{\phi}$ which maximizes the log likelihood $l(\phi; x, v)$ based on the observations. However, if l has multiple maxima, there is no guarantee that $\{\hat{\phi}_j\}$ will converge to the global maximum, and some rather artificial examples can be constructed in which the sequence $\{\hat{\phi}_j\}$ converges to a saddle point of l; see also Exercise 11.5.

11.2   Some illustrations of the EM algorithm

We now consider some examples of the use of the EM algorithm.

(i)   *Exponentially distributed survival times*

The maximum likelihood estimator $\hat{\rho}$, based on a censored sample from the exponential distribution with parameter ρ, was derived explicitly in Chapter 3 as

$$\hat{\rho} = d / \sum x_i;\qquad (11.4)$$

see equation (3.13). Iterative procedures are, therefore, unnecessary. However, it is of interest to see how the EM algorithm works in this very simple situation.

The log likelihood $l_0(\rho'; T)$ based on an uncensored exponential sample $T = (T_1, \ldots, T_n)$ would be

$$l_0(\rho'; T) = n \log \rho' - \rho' \sum_{i=1}^{n} T_i.$$

The function $Q(\rho', \rho)$ defined at (11.3) becomes

$$Q(\rho', \rho) = n \log \rho' - \rho' \sum_{i=1}^{n} E(T_i | x_i, v_i; \rho). \tag{11.5}$$

For an uncensored observation $(v_i = 1)$, we have $T_i = x_i$. For a censored observation $(v_i = 0)$, $E(T_i | x_i, v_i; \rho) = x_i + \rho^{-1}$ by the lack of memory property. Thus,

$$E(T_i | x_i, v_i; \rho) = x_i + (1 - v_i)/\rho.$$

Substitution in (11.5) gives

$$Q(\rho', \rho) = n \log \rho' - \rho' \left(\sum_{i=1}^{n} x_i + \frac{n-d}{\rho} \right). \tag{11.6}$$

This calculation defines the E step of the algorithm. To carry out the M step, we must maximize $Q(\rho', \rho)$ in ρ', for fixed ρ. This is easily done,

Table 11.1 EM algorithm applied to the
leukaemia data (treated group) assuming
exponential distributions

Iteration	$\hat{\rho}_j$	$l(\hat{\rho})$
0	0.05850	− 46.5501
1	0.03723	− 42.9808
2	0.03082	− 42.3808
3	0.02806	− 42.2343
4	0.02670	− 42.1930
5	0.02597	− 42.1806
6	0.02558	− 42.1767
7	0.02536	− 42.1755
8	0.02523	− 42.1751
9	0.02516	− 42.1749
10	0.02512	− 42.1749
11	0.02510	− 42.1749
12	0.02509	− 42.1749
13	0.02508	− 42.1749
14	0.02508	− 42.1749
15	0.02507	− 42.1749
$\hat{\rho}$	0.02507	− 42.1749

Initial estimate $\hat{\rho}_0 = 21/359 = 0.05850$

and the two steps may be combined to give the formula

$$\hat{\rho}_{j+1} = n\left(\sum_{i=1}^{n} x_i + \frac{n-d}{\hat{\rho}_j}\right)^{-1} \tag{11.7}$$

for calculating the $(j+1)$th approximation to $\hat{\rho}$ from the jth approximation. The substitution $\hat{\mu}_j = 1/\hat{\rho}_j$ allows (11.7) to be expressed as

$$\hat{\mu}_{j+1} = n^{-1}\sum x_i + (1-d/n)\hat{\mu}_j,$$

which shows immediately that $\hat{\mu}_j \to \sum x_i/d = \hat{\rho}^{-1} = \hat{\mu}$ say. In fact, since

$$\hat{\mu}_{j+1} - \hat{\mu} = (1-d/n)(\hat{\mu}_j - \hat{\mu}),$$

the convergence on the μ-scale is exactly geometric, with rate depending on the proportion of censored observations.

Table 11.1 shows the results of applying the EM algorithm to the leukaemia data (treatment group) given in Table 1.1 and discussed in Chapter 3. For each iteration, the table gives the current estimate of the parameter ρ, and the log likelihood. Because of the high rate of censoring, the convergence is quite slow.

(ii)   The gamma distribution

As a slightly more complex example, consider the estimation of the parameters (κ, ρ) of the gamma distribution, with density

$$f(t;\kappa,\rho) = \frac{\rho(\rho t)^{\kappa-1}e^{-\rho t}}{\Gamma(\kappa)},$$

from a censored sample (x, v). The log likelihood for an uncensored sample would be

$$l_0(\phi; T) = n\kappa \log \rho - n(\kappa-1)\sum \log T_i - \rho \sum T_i - n \log \Gamma(\kappa), \tag{11.8}$$

which depends on the $\{T_i\}$ only through the statistics $S_1 = \sum T_i$ and $S_2 = \sum \log T_i$, which are, therefore, jointly sufficient for $\phi = (\kappa, \rho)$. In Exercise 3.5 it was shown that the maximum likelihood estimate $\hat{\kappa}$ of κ is the solution of the equation

$$\psi(\kappa) - \log \kappa - \log R = 0, \tag{11.9}$$

where $\psi(\kappa)$ denotes the digamma function, $\psi(\kappa) = d \log \Gamma(\kappa)/d\kappa$, and

$$R = n\exp(n^{-1}S_2)/S_1 \qquad (11.10)$$

is the ratio of the geometric to the arithmetic mean of the sample. Then $\hat{\rho}$ is obtained as $\hat{\rho} = n\hat{\kappa}/S_1$.

The log likelihood $l(\phi:x,v)$ from a censored sample, involves the incomplete gamma function, and has no useful sufficient statistics. To see how the EM algorithm fares here, note that evaluation of $Q(\phi',\phi)$ involves the calculation of the conditional expectations

$$E(S_1|x,v;\phi), \qquad E(S_2|x,v;\phi) \qquad (11.11)$$

of the sufficient statistics S_1 and S_2 given the data (x, v) for the current parameter estimates ϕ. These expectations also involve the incomplete gamma function and its derivative with respect to the index parameter, so that there is no great saving in computation, over, say, a Newton–Raphson procedure for the maximization of $l(\phi; x, v)$. The maximization step of the EM algorithm is more straightforward, the root $\hat{\kappa}$ of (11.9) being required when S_1 and S_2 are replaced by their conditional expectations (11.11).

Properties of the EM algorithm as applied to general exponential families, of which the gamma distribution is an example, are derived in a little more detail in Section 11.4. We illustrate the present discussion by applying the method to some data on the survival times for 22 patients with bile duct cancer treated by radiation–drug therapy; see Table 11.2. The coefficient of variation of the 19 uncensored times is 0.66, suggesting that a gamma distribution with $\kappa \approx 1/(0.66)^2 \approx 2$ might provide a reasonable fit. The algorithm gives maximum likelihood estimates $\hat{\rho} = 0.008379$, $\hat{\kappa} = 2.3648$.

Table 11.2 *Use of the EM algorithm to fit a gamma distribution. Survival times (days) of 22 patients with bile duct cancer treated with combined drug and radiation therapy (data of Fleming et al., 1980)*

(a) *Data*

Uncensored times	30,	67,	95,	148,	170,	171,	176,	193,	200,	221,
	243,	261,	262,	263,	399,	414,	446,	464,	777	
Censored times		79,	82,	446						

Table 11.2 (*contd.*)

(b) *Computations*

Iteration	Parameter estimates		Log likelihood	$E(T\|T > c)$*	$E(\log T\|T > c)$*
	$\hat{\rho}$	$\hat{\kappa}$	l		
1[†]	0.008670	2.2096	−123.932	280.48	5.4806
				282.20	5.4900
				590.89	6.3583
2	0.008508	2.3797	−123.665	300.47	5.5486
				301.97	5.5566
				599.27	6.3707
3	0.008403	2.3687	−123.663	302.70	5.5550
				304.20	5.5629
				601.21	6.3735
4	0.008383	2.3655	−123.663	303.03	5.5558
				304.53	5.5637
				601.55	6.3740
5	0.008379	2.3649	−123.663	303.08	5.5560
				304.58	5.5639
				601.61	6.3741
6	0.008379	2.3648	−123.663	303.09	5.5560
				304.59	5.5639
				601.62	6.3741

* For the three censored times.

11.3 Behaviour of the algorithm: self-consistency

Although a detailed discussion of the properties of the EM algorithm would be inappropriate in a book such as this, one important and easily derived result goes far to establish the usefulness of the technique. The log likelihood $l(\phi;x,v)$ is never decreased at any iteration of the algorithm.

To see this, note first that

$$l(\phi';x,v) = l_0(\phi';T) - l_1(\phi';T|x,v), \qquad (11.12)$$

where ϕ' is any value of the parameter ϕ, and l_1 is the log likelihood from the conditional distribution of T given (x,v). Since $l(\phi')$ depends on T only through the observations (x, v), it equals its expectation with respect to *any* conditional distribution of T given (x, v). Thus we may write

$$l(\phi') = Q(\phi',\phi) - R(\phi',\phi), \qquad (11.13)$$

where

$$R(\phi';\phi) = E[l_1(\phi';T)|x,v;\phi].$$

It follows from Jensen's inequality and the concavity of the logarithm function that if $g(u;\phi)$ is the density, or discrete probability function, of a random variable U, then

$$E\{\log[g(U;\phi')];\phi\} \leqslant E\{\log[g(U;\phi)];\phi\}.$$

The expectation of the log likelihood is maximized at the true value of the parameter. Application of this inequality to the conditional log likelihood of T given (x, v) shows that $R(\phi',\phi) \leqslant R(\phi,\phi)$ for any ϕ, ϕ'.

From (11.13) we have that

$$l(\phi') - l(\phi) = [Q(\phi',\phi) - Q(\phi,\phi)] - [R(\phi',\phi) - R(\phi,\phi)]. \qquad (11.14)$$

If ϕ' is chosen, as at the M step of the EM algorithm, to maximize $Q(\phi',\phi)$ in ϕ' for the given previous value ϕ, then it follows that $l(\phi') - l(\phi) \geqslant 0$, so that the iteration can never decrease the log likelihood.

An immediate consequence is that the maximum likelihood estimator $\hat{\phi}$ must satisfy the self-consistency condition

$$Q(\phi',\hat{\phi}) \leqslant Q(\hat{\phi},\hat{\phi}), \qquad (11.15)$$

since if we start at $\phi = \hat{\phi}$, the log likelihood cannot be increased by a step of the EM algorithm, or in any other way. The condition (11.15), or its differential form, that $\hat{\phi}$ is a solution of the equation in ϕ,

$$\left[\frac{\partial}{\partial \phi'} Q(\phi', \phi) \right]_{\phi' = \phi} = 0 \qquad (11.16)$$

is often more useful than the algorithm itself, whose convergence may be slow. Thus, in the first example of the preceding section, equation (11.6) gives

$$\frac{\partial}{\partial \rho'} Q(\rho', \rho) = \frac{n}{\rho'} - \left(\sum_{i=1}^{n} x_i + \frac{n-d}{\rho} \right),$$

and so (11.16) gives

$$\frac{n}{\rho} - \left(\sum_{i=1}^{n} x_i + \frac{n-d}{\rho} \right) = 0,$$

or $\hat{\rho} = d / \sum x_i$, for the maximum likelihood estimator $\hat{\rho}$. Equations (11.15) and (11.16) are conditions for a fixed point of the EM algorithm. We have shown that a maximum likelihood estimator must be a fixed point of the algorithm, but there is no guarantee that it is the only one.

We conclude this section with a brief comparison of the EM algorithm with its main competitor, direct maximization of $l(\phi; x, v)$ by the Newton–Raphson procedure. The latter method requires calculation and inversion of the matrix of second derivatives, a time-consuming procedure if ϕ has large dimension. Divergence is much more frequent, as a full Newton–Raphson step will not necessarily increase the log likelihood. As against this, the Newton–Raphson procedure usually converges rapidly when it does converge, in particular when the log likelihood is well approximated by a quadratic function, and the inverse of the matrix of second derivatives is often needed for the estimation of standard errors.

It is important to note that the second derivatives of the log likelihood $l_0(\hat{\phi}; T)$ or of its expectation $Q(\hat{\phi}, \hat{\phi})$ at the maximum likelihood estimator do not provide valid estimates of standard error. If the EM algorithm is used to find $\hat{\phi}$, a separate evaluation of the second derivatives of $l(\phi; x, v)$ at $\phi = \hat{\phi}$ is needed. A method for estimating these standard errors without evaluating l, and of accelerating the convergence of the algorithm in the neighbourhood of $\hat{\phi}$, is indicated in Exercises 11.3 and 11.4.

11.4 Exponential families

Suppose that the density of T_i is a regular member of the full exponential family in its natural parameterization, i.e.

$$f_i(t;\phi) = \exp[\phi^T S_i(t) + A_i(t) + B_i(\phi)], \qquad (11.17)$$

where, for each i, $S_i(t)$ is a $p \times 1$ vector of linearly independent functions of t, $A_i(t)$ is a scalar function of t and $B_i(\phi)$ is a scalar function of the parameter vector ϕ. Note that by writing $S_i(t) = S(t, z_i)$ we may express (11.17) in a form which allows the distribution of survival time to depend on a vector z of explanatory variables.

The importance of the exponential family in the general theory of statistical inference arises because the likelihood from an (uncensored) sample of size n possesses a sufficient statistic of fixed dimension,

$$S = \sum_{i=1}^{n} S_i(T_i).$$

By differentiation under the integral of the total probability, it follows that

$$E(S_i) = -\partial B_i(\phi)/\partial\phi, \qquad (11.18)$$

where $\partial/\partial\phi$ denotes the gradient with respect to the vector ϕ. With $B(\phi) = \sum B_i(\phi)$, the likelihood equations may be written

$$0 = \frac{\partial l_0(\phi)}{\partial\phi} = S - E(S; \phi), \qquad (11.19)$$

so that solution of the likelihood equations involves equating each component of S to its expectation. It is easily shown also that $l_0(\phi)$ is concave in ϕ.

For a censored sample (x, d), the function $Q(\phi', \phi)$ defined at (11.3) takes the form

$$Q(\phi', \phi) = \phi'^T E(S|x, v; \phi) + B(\phi'),$$

up to an additive function not depending on ϕ'. Maximization of Q in ϕ' is achieved by setting $\partial/\partial\phi'[Q(\phi', \phi)] = 0$, giving

$$E(S|x, v; \phi) = -\partial B(\phi')/\partial\phi' = E(S; \phi'), \qquad (11.20)$$

as an equation in ϕ', for given ϕ. Note that (11.20) has the same form as (11.19). The two steps of the EM algorithm can be expressed as

(i) E step: calculate $S' = E(S|x, v; \phi)$;
(ii) M step: obtain ϕ' as the solution of $E(S; \phi') = S'$.

In the expectation step, we have

$$E(S_i|x_i, v_i) = v_i S_i(x_i) + (1 - v_i) E[S_i(T_i)|T_i > x_i].$$

Thus $S_i(x_i)$ is left unchanged for an uncensored observation, but replaced by its conditional expectation given $T_i > x_i$ for an observation censored at x_i. Calculation of the expectation

$$E[S_i(T_i)|T_i > x_i] = \frac{1}{\mathcal{F}_i(x_i; \phi)} \int_{x_i}^{\infty} S_i(t) f(t; \phi) dt$$

is usually the more difficult step in the implementation.

A re-expression of the log likelihood $l(\phi; x, v)$ gives some insight into the nature of the algorithm. From (11.12),

$$l(\phi; x, v) = \sum_{i=1}^{n} \log f_i(T_i; \phi)$$

$$- \sum_{i=1}^{n} (1 - v_i) \log[f_i(T_i; \phi)/\mathcal{F}_i(x_i; \phi)]. \quad (11.21)$$

Now, $f_i(t; \phi)/\mathcal{F}_i(x; \phi)$ is the conditional density of T_i given $T_i > x$. It has the same form as the unconditional density of T_i given in (11.17), except that the range of the density is restricted to $t > x$ and the normalizing constant $B_i(\phi, x)$ now depends on x. As before,

$$\frac{\partial}{\partial \phi} \log\left(\frac{f_i(T_i; \phi)}{\mathcal{F}_i(x_i; \phi)}\right) = S_i(T_i) - E[S_i(T_i)|T_i > x_i]. \quad (11.22)$$

Combining equations (11.21) and (11.22), we obtain

$$\frac{\partial l(\phi; x, v)}{\partial \phi} = S - E(S; \phi)$$

$$- \sum_{i=1}^{n} (1 - v_i)\{S_i(T_i) - E[S_i(T_i)|T_i > x_i]\}$$

$$= E(S|x, v; \phi) - E(S; \phi), \quad (11.23)$$

the difference between the conditional and unconditional expectations of the sufficient statistic. Thus, if ϕ is a fixed point of the algorithm, so that $\phi' = \phi$ in (11.20), then ϕ is also a solution of the likelihood equation $\partial l/\partial \phi = 0$.

11.5 Application to grouped multinomial likelihoods

We now consider a fairly general formulation of the self-consistency concept for likelihoods obtained from grouped multinomial data. This approach gives another derivation of the product-limit estimator discussed in Chapter 4, and allows this to be extended to the case where survival times are subject to more general patterns of censoring and grouping than have been considered so far in this book. It also encompasses the problem discussed in Chapter 10, of estimating a bivariate survival distribution from data that are censored in either or both components.

Suppose that a multinomial experiment has p possible outcomes with probabilities $\pi_j(j = 1, \ldots, p)$, with $\sum \pi_j = 1$, but that in the ith replication $(i = 1, \ldots, n)$ the result of the experiment is observed only to lie in the set $S_i \subset \{1, \ldots, p\}$. Then the overall log likelihood is

$$l(\pi) = \sum_{i=1}^{n} \log\left(\sum_{j \in S_i} \pi_j \right). \tag{11.24}$$

It is easy to show that $l(\pi)$ is concave. For the gradient is

$$\frac{\partial l}{\partial \pi} = \sum_i \frac{g_i}{|S_i|}, \tag{11.25}$$

where g_i is a $p \times 1$ vector with components $g_{ij} = 1$ if $j \in S_i, g_{ij} = 0$ otherwise, and

$$|S_i| = \sum_{j \in S_i} \pi_j.$$

The matrix of second derivatives of l is $-I$, where

$$I = \sum_i g_i g_i^{\mathrm{T}}/|S_i|^2 = J^{\mathrm{T}} J, \tag{11.26}$$

and J is the $n \times p$ matrix with ith row $g_i^{\mathrm{T}}/|S_i|$. This shows that the matrix I is positive semidefinite and that the function $l(\pi)$ is concave. Since the simplex $\{\pi; \sum \pi_j = 1\}$ of admissible parameter values is a closed convex set, the values of π for which the maximum is achieved also form a convex set, and l can have no other stationary values. The maximum need not, however, be unique.

We now consider the implementation of the EM algorithm to this problem. As π is not the so-called 'canonical' parameter vector for the multinomial distribution, so that the theory of Section 11.4 does not directly apply, it is convenient to work instead from the general formulation in Section 11.3. The full data here would consist of the multinomial frequencies K_j of each outcome $(j = 1, \ldots, p)$ that would be observed if there was no grouping. The actual observations consist of the sets S_i associated with each replication.

The full likelihood takes the multinomial form

$$l_0(\phi ; K)' = \sum_{j=1}^{p-1} K_j \log \pi_j$$
$$+ K_p \log(1 - \pi_1 - \ldots - \pi_{p-1}), \qquad (11.27)$$

where we have substituted for π_p to avoid the constraint $\sum \pi_j = 1$. The function $Q(\pi', \pi)$ defined at (11.3) takes the form

$$Q(\pi', \pi) = \sum_{j=1}^{p-1} E(K_j | S, \pi) \log \pi_j'$$
$$+ E(K_p | S, \pi) \log(1 - \pi_1' - \ldots - \pi_{p-1}'),$$

where $S = \{S_1, S_2, \ldots, S_n\}$ denotes the observed data. Both steps of the algorithm are straightforward. For the expectation step we have that

$$E(K_j | S, \pi) = \sum_{i=1}^{n} \frac{g_{ij} \pi_j}{|S_i|}, \qquad (11.28)$$

and for the maximization step,

$$\pi_j' = E(K_j | S, \pi) / n. \qquad (11.29)$$

The self-consistency equations become

$$\sum_{i=1}^{n} \frac{g_{ij} \pi_j}{|S_i|} = n \pi_j \qquad (j = 1, \ldots, p-1). \qquad (11.30)$$

Cancellation of the π_j is not in general permissible, as we may have $|S_i| = \pi_j$ for some i.

Despite the concavity of the log likelihood, convergence of the EM algorithm to a maximum likelihood estimate is not guaranteed, due to possible anomalous behaviour on the boundaries of the parameter space (Exercise 11.5).

As an illustration of the use of the algorithm, we consider again the estimation of a distribution from a single-sample subject to right censoring. Here, using the notation of Chapter 4, the d_j observed failures at the possible failure time a_j each have $S_i = \{a_j\}$, whereas the m_j censorings at a_j have $S_i = \{a_{j+1}, \ldots, a_p\}$. Note that, as before, we assume that an individual i censored at a_j could have been observed to fail at a_j, so that a_j is excluded from S_i.

The self-consistency equations (11.30) become

$$n\pi_j = d_j + \sum_{i<j} \frac{m_i \pi_j}{1 - \pi_1 - \pi_2 - \ldots - \pi_i}. \tag{11.31}$$

It is easily shown by mathematical induction on j that the solution of (11.31) is

$$\hat{\pi}_j = \left[\prod_{i=1}^{j-1} \left(1 - \frac{d_i}{r_i} \right) \right] \frac{d_j}{r_j}, \tag{11.32}$$

where, as before,

$$r_j = n - \sum_{i<j} (d_i + m_i)$$

is the number of subjects at risk at a_j. This corresponds to the product-limit estimator for the survivor function

$$\hat{\mathscr{F}}(t) = 1 - \sum^{(t)} \hat{\pi}_j$$
$$= \prod^{(t)} (1 - d_j/r_j)$$

derived in (4.4).

11.6 Other schemes of observation

Throughout the discussion in this book, allowance has been made for possible right censoring. Indeed any method of statistical analysis for failure data that does not fairly readily accommodate such censoring is of rather restricted usefulness. There are, however, other possible restrictions on the observation of failure time and we conclude by describing briefly how these can be dealt with.

Left truncation arises when individuals come under observation only some known time after the natural time origin of the phenomenon under study. That is, had the individual failed before the truncation time in question, that individual would not have been recorded. Therefore, in particular, any contribution to the likelihood must be conditional on the truncation limit having been exceeded.

If for the ith individual the left truncation limit is t_i', possibly zero, and the individual either fails at time t_i or is right censored at time c_i, the contribution to the likelihood for a homogeneous sample of individuals is either

$$f(t_i;\phi)/\mathscr{F}(t_i';\phi) \qquad \text{or} \qquad \mathscr{F}(c_i;\phi)/\mathscr{F}(t_i';\phi) \qquad (11.33)$$

and this leads directly to maximum likelihood fitting of any parametric model. Explanatory variables can be introduced.

In particular, as is clear on general grounds, the discussion for exponential distributions is particularly simple. With $\mathscr{F}(t;\rho) = e^{-\rho t}$, the factors (11.33) become either $\rho e^{-\rho(t_i-t_i')}$ or $e^{-\rho(c_i-t_i')}$ and, so far as the likelihood analysis is concerned, only the number of failures and total time under study are relevant. For other parametric distributions, numerical solution of the maximum likelihood equations is in general needed, but no new principle is involved.

Censoring on the left arises if observation does not start immediately and some individuals have already failed before it does. For these individuals the fact that they have failed, but not their failure time, is known. This can happen if the 'failure time' of interest is the first time at which some characteristic level has been achieved by an increasing function, and the level may have been reached for some individuals before observation begins. In general, an individual whose failure time is left censored at c contributes a term $F(c;\phi)$ to the parametric likelihood function.

The EM algorithm as described in the previous section can cope with the nonparametric estimation of a distribution function from survival times subject to both left and right censoring. Of course, if only left censoring is present, the usual product-limit estimator may be applied after a change in sign.

For truncated data, a slight modification of the algorithm is needed, but the important special case of left truncation combined with right censoring has a simple explicit solution. In fact, equation (11.32) still holds, provided that the definition of r_j is modified so that it counts only those individuals who enter observation before a_j.

In some contexts, in particular in reliability studies and some sociological investigations, where there is a stationary process of intervals in time on each of a population of individuals, other schemes of observation are possible, in particular the following:

(i) a cross-sectional sample of a population may be taken at a particular instant, and the complete failure times obtained of the individuals captured in the sample;

(ii) a sample may be obtained as in (i) but, instead of recording the complete failure time, we record either the remaining lifetime (forward recurrence time) measured forward from the instant of sampling to failure or, alternatively, the corresponding backward recurrence time.

Quite often, if $f(x; \phi)$ and $\mathscr{F}(x; \phi)$ are the density and survivor function of failure time, of mean μ, the procedures generate observations with (i) the length-biased density, $xf(x;\phi)/\mu$; and (ii) the equilibrium recurrence-time density, $\mathscr{F}(x;\phi)/\mu$. These densities certainly apply to the cross-sectional sampling of a stationary point process, with interval density $f(x; \phi)$. Having obtained data from the densities (i) and (ii), we can write down likelihood functions for parametric estimation. Cox (1967) has examined the asymptotic efficiency of estimation relative to direct observation of failure times. For nonparametric estimation from length-biased sampling, see Vardi (1982a, b), and for related applications in geometric probability, Laslett (1982).

Bibliographic notes, 11

The methods devised by Yates (1933) and Healy and Westmacott (1956) for dealing with missing values in designed experiments are perhaps the earliest examples of the self-consistency principle. Efron (1967) defined the term and was the first to apply the concept to censored data. The term EM algorithm was introduced by Dempster *et al.* (1977) in a comprehensive paper which surveyed earlier applications and introduced many new ones. See Wu (1983) for convergence results. Orchard and Woodbury (1972) coined the term missing information principle.

Burridge (1981a, 1982) has given conditions for concavity of log likelihood functions from grouped and censored data. See Turnbull (1974, 1976) for nonparametric estimation from data subject to arbitrary patterns of grouping, censoring and truncation. Miller (1976) and Buckley and James (1979) have appealed to the concept of self-consistency to derive generalized least-squares estimates for regression models with censored data, and Miller and Halpern (1982) apply these procedures to analyse post-transplant survival in heart transplant patients; see also Section 8.9.

Further results and exercises, 11

11.1. Suppose that the log failure times $Y_i = \log T_i$ follow the standard linear model, $Y \sim N(\beta^T z, \sigma^2 I^*)$, where z is a $p \times n$ design matrix and I^* is the $n \times n$ identity matrix. Show that the expectation step of the EM algorithm involves calculation of the conditional expectations $E(Y_i | Y_i > \log c_i)$ and $E(Y_i^2 | Y_i > \log c_i)$, but that only the former values are needed to update the regression parameters β. Express both conditional expectations in terms of the normal cumulative distribution function and hence show how to implement the algorithm.

Would you expect the maximum likelihood estimate of β to be affected if σ^2 were specified in advance?

[Glasser, 1965; Schmee and Hahn, 1979; Wolynetz, 1979; Aitkin, 1981]

11.2. Show that if, in a regular estimation problem, the EM algorithm converges to an interior point ϕ_0 of the parameter space, then ϕ_0 must be a stationary point of the log likelihood. [Hint: Consider the gradient of $R(\phi', \phi)$ in ϕ' at $\phi = \phi_0$.]

11.3. Let $U_0(\phi), U(\phi), I_0(\phi)$ and $I(\phi)$ denote the gradients and observed information matrices from $l_0(\phi:T)$ and $l(\phi;x,v)$ respectively. Let $\hat{\phi}$ be the maximum likelihood estimator of ϕ, i.e. from l. Show that, in a regular estimation problem,

(a) $U(\phi) = E[U_0(\phi)|x,v]$,
(b) $U(\hat{\phi}) = 0$,
(c) $I(\phi) = E[I_0(\phi)|x,v] - E[U_0(\phi)U_0^T(\phi)|x,v] + U(\phi)U^T(\phi)$,

where all expectations are taken with respect to the conditional distribution of T given (x,v).

Show how these equations may be used to obtain an estimate of the standard error of $\hat{\phi}$. Does this method lead to any computational saving over a direct evaluation of $I(\hat{\phi})$?

[Louis, 1982]

11.4. (Continuation) Writing $I = I(\hat{\phi})$ and $I_0 = E[I_0(\hat{\phi})|x,v]$ and $\delta^{(j)} = \phi^{(j)} - \phi^{(j-1)}$ for the difference between the $(j-1)$th and jth approximations to $\hat{\phi}$ from the EM algorithm, show that, neglecting second-order terms in $\phi^{(j)} - \hat{\phi}$,

$$\delta^{(j)} = J\delta^{(j-1)},$$

where

$$(I^* - J)^{-1} = I_0 I^{-1}.$$

Show how this result may be used to 'accelerate' the convergence of the algorithm.

[Sundberg, 1974, 1976; Louis, 1982]

11.5. In the notation of Section 11.5, suppose that in the three replications of a multinomial experiment with three outcomes, the observed data are $S_1 = \{2, 3\}$, $S_2 = \{3, 1\}$ and $S_3 = \{1, 2\}$. Show that there is a unique maximum likelihood estimate of π, namely $\hat{\pi}_1 = \hat{\pi}_2 = \hat{\pi}_3 = \frac{1}{3}$.

Show, however, that the EM algorithm converges to the point $\pi_1 = 0$, $\pi_2 = \pi_3 = \frac{1}{2}$ from any starting point, with $\pi_1 = 0$, $\pi_2 \neq 0$, $\pi_3 \neq 0$.

What are the other possible limit points of the EM algorithm here?

11.6. Consider a number of identical systems each consisting of two components, A and B. Each system can function if either component can. Let π_{ij} denote the probability that component A fails in year i, component B in year j ($i, j = 1, 2, 3$) and suppose for definiteness that no component can last longer than three years. Suppose that initially 100 such systems are tested, with outcomes, in an obvious notation, $k_{11} = 10$, $k_{12} = 20$, $k_{13} = 10$, $k_{21} = 15$, $k_{22} = 8$, $k_{23} = 16$, $k_{31} = 5$, $k_{32} = 12$ and $k_{33} = 4$. Suppose that in a second experiment, a further 1000 identical systems are tested, but that in this second experiment observation ceases at the end of the first year, when 110 systems have failed in both A and B, 325 have failed in A only, 275 in B only and 290 are still fully operational. Use the EM algorithm to derive the maximum likelihood estimates of π_{ij}.

[Campbell, 1981]

11.7. (Additive hazards) Suppose that the hazard function for a single homogeneous sample is $h_0(t) + \theta h_1(t)$ where $h_0(t)$ and $h_1(t)$ are specified functions of t and $\theta > 0$ is a parameter to be estimated. Derive the likelihood equation for θ (a) directly and (b) by the EM algorithm. [Hint: suppose there are two types of failure, with (independent) hazard rates $h_0(t)$ and $\theta h_1(t)$, but that the type of failure is not observed.]

11.8. A family of survivor functions $\mathscr{F}(t; \phi)$ is said to be closed under truncation if $\mathscr{F}(t;\phi)/\mathscr{F}(t';\phi) = \mathscr{F}(t - t';\phi')$, where ϕ' may depend on ϕ and t'. Which of the families introduced in Chapter 2 are closed under truncation?

11.9. In animal carcinogenesis experiments, a tumour is *fatal* if it causes the death of the animal, *incidental* if it does not itself cause death but can be detected only when the animal is sacrificed or dies from other causes, and *visible* if its existence can be detected during the life of the animal. Show that times to fatal or to visible tumours are either observed or right censored, but that times to incidental tumours are all either right censored or left censored. Discuss the implications for analysis.

[Peto *et al.*, 1980]

References

Aalen, O. (1976). Nonparametric inference in connection with multiple decrement models. *Scand. J. Statist.*, **3**, 15–27.

Aalen, O. (1978). Nonparametric inference for a family of counting processes. *Ann. Statist.*, **6**, 701–26.

Aitkin, M. (1981). A note on the regression analysis of censored data. *Technometrics*, **23**, 161–3.

Aitkin, M. and Clayton, D. (1980). The fitting of exponential, Weibull and extreme value distributions to complex censored survival data using GLIM. *Appl. Statist.*, **29**, 156–63.

Aitkin, M., Laird, N. and Francis, B. (1983). A reanalysis of the Stanford heart transplant data (with discussion). *J. Am. Statist. Assoc.*, **78**, 264–92.

Andersen, P.K., Borgan, O., Gill, R.D. and Keiding, N. (1982). Linear nonparametric tests for comparison of counting processes with applications to censored survival data. *Int. Statist. Rev.*, **50**, 219–58.

Andersen, P.K. and Gill, R.D. (1982). Cox's regression model for counting processes: a large sample study. *Ann. Statist.*, **10**, 1100–20.

Anderson, J.A. and Senthilselvan, A. (1980). Smooth estimates for the hazard function. *J. R. Statist. Soc.*, B **42**, 322–7.

Aranda-Ordaz, F.J. (1980). Transformations to additivity for binary data. Unpublished Ph.D. thesis, Imperial College, London.

Armitage, P. (1959). The comparison of survival curves (with discussion). *J. R. Statist. Soc.*, A **122**, 279–300.

Barlow, R.E. and Proschan, F. (1965). *Mathematical Theory of Reliability.* New York: Wiley.

Barlow, R.E. and Proschan, F. (1975). *Statistical Theory of Reliability and Life Testing.* New York: Holt, Rinehart and Winston.

Barndorff-Nielsen, O. (1980). Conditionality resolutions. *Biometrika*, **67**, 293–310.

Barndorff-Nielsen, O. (1983). On a formula for the distribution of the maximum likelihood estimator. *Biometrika*, **70**, 343–65.

Barndorff-Nielsen, O. and Cox, D.R. (1984). Bartlett adjustments to the likelihood ratio statistic and the distribution of the maximum likelihood estimator. *J. R. Statist. Soc.*, B **46**, to appear.

Bartholomew, D.J. (1957). A problem in life testing. *J. Am. Statist. Assoc.*, **52**, 350–5.

Bartholomew, D.J. (1963). The sampling distribution of an estimate arising in life testing. *Technometrics*, **5**, 361–74.

Bennett, S. (1983). Analysis of survival data by the proportional odds model. *Statist. Med.*, **2**, 273–8

Bernardo, J.M. (1976). Psi (digamma) function. Algorithm AS103. *Appl. Statist.*, **25**, 315–17.

Bernoulli, D. (1760). Essai d'une nouvelle analyse de la mortalité causée par la petite Vérole, et des avantages de l'Inoculation pour la prevenir. *Mem. L'Acad. R. Sci.*, **1760**, 1–45.

Billingsley, P. (1968). *Convergence of Probability Measures*. New York: Wiley.

Billingsley, P. (1971). *Weak Convergence of Measures*. Philadelphia: SIAM.

Boag, J.W. (1949). Maximum likelihood estimates of the proportion of patients cured by cancer therapy (with discussion). *J. R. Statist. Soc.*, B **11**, 15–53.

Böhmer, P.E. (1912). Theorie der unabhängigen Wahrscheinlichkeiten *Rapports Mémoires et Procès-verbaux de Septième Congrès International d'Actuaires*, Amsterdam, vol. 2, pp. 327–43.

Borgefors, G. and Hjorth, U. (1981). Comparison of parametric models for estimating maintenance times from small samples. *IEEE Trans. Rel.*, **R-30**, 375–80.

Bradford Hill, A. (1977). *A Short Textbook of Medical Statistics*. London: Hodder and Stoughton.

Breslow, N.E. (1970). A generalized Kruskal–Wallis test for comparing *K* samples subject to unequal patterns of censorship. *Biometrika*, **57**, 579–94.

Breslow, N.E. (1972). Contribution to discussion of paper by D.R. Cox. *J. R. Statist. Soc.*, B **34**, 216–17.

Breslow, N.E. (1974). Covariance analysis of censored survival data. *Biometrics*, **30**, 89–100.

Breslow, N.E. (1975). Analysis of survival data under the proportional hazards model. *Int. Statist. Rev.*, **43**, 55–8.

Breslow, N.E. (1977). Some statistical models useful in the study of occupational mortality, in *Environmental Health : Quantitative Methods* (ed. A. Whittemore). Philadelphia: SIAM. pp. 88–102.

Breslow, N.E. and Crowley, J. (1974). A large sample study of the life table and product limit estimates under random censorship. *Ann. Statist.*, **2**, 437–53.

Breslow, N.E. and Crowley, J. (1981). Discussion of paper by D. Oakes. *Int. Statist. Rev.*, **49**, 255–7.

Breslow, N.E. and Day, N.E. (1980). *Statistical Methods in Cancer Research*, vol. 1, *The Analysis of Case-Control Studies*. Lyon: IARC.

Breslow, N.E., Lubin, J.H., Marek, P. and Langholtz, B. (1983). Multiplicative models and cohort analysis. *J. Am. Statist. Assoc.*, **78**, 1–12.

Brown, B.W., Hollander, M. and Korwar, R.M. (1974). Nonparametric tests of independence for censored data, with applications to heart transplant studies, in *Reliability and Biometry. Statistical Analysis of Life Length* (eds F. Proschan and R.J. Serfling). Philadelphia: SIAM. pp. 327–53.

Brown, M.B. (1984). On the choice of variance for the log rank test. *Biometrika*, **71**, 65–74.

Buckley, J. and James, I. (1979). Linear regression with censored data. *Biometrika*, **66**, 429–36.

Burridge, J. (1981a). A note on maximum likelihood estimation for regression models using grouped data. *J. R. Statist. Soc.*, B **43**, 41–5.

Burridge, J. (1981b). Empirical Bayes analysis of survival time data. *J. R. Statist. Soc.*, B **43**, 65–75.

Burridge, J. (1981c). Statistical analysis of grouped lifetime data in a changing environment. Unpublished Ph.D. thesis, Imperial College, London.

Burridge, J. (1982). Some unimodality properties of likelihoods derived from grouped data. *Biometrika*, **69**, 145–51.

Campbell, G. (1981). Nonparametric bivariate estimation with randomly censored data. *Biometrika*, **68**, 417–22.

Clausius, R. (1858). Ueber die Mittlere Länge der Wege. *Ann. Phys. Lpzg*, **105**, 239–58.

Clayton, D.G. (1974). Some odds ratio statistics for the analysis of ordered categorical data. *Biometrika*, **61**, 525–31.

Clayton, D.G. (1978). A model for association in bivariate life tables and its application in epidemiological studies of familial tendency in chronic disease incidence. *Biometrika*, **65**, 141–51.

Cochran, W.G. (1954). Some methods for strengthening the common χ^2 tests. *Biometrics*, **10**, 417–51.

Cornfield, J. and Detre, K. (1977). Bayesian life table analysis. *J. R. Statist. Soc.*, B **39**, 86–94.

Cox, D.R. (1953). Some simple approximate tests for Poisson variates. *Biometrika*, **40**, 354–60.

Cox, D.R. (1959). The analysis of exponentially distributed life-times with two types of failure. *J. R. Statist. Soc.*, B, **21**, 411–21.

Cox, D.R. (1962). *Renewal Theory.* London: Methuen.

Cox, D.R. (1967). Some sampling problems in technology, in *New Developments in Survey Sampling* (eds N.L. Johnson and H. Smith). New York: Wiley, pp. 506–27.

Cox, D.R. (1970). *Analysis of binary data.* London: Chapman and Hall.

Cox, D.R. (1972). Regression models and life-tables (with discussion). *J. R. Statist. Soc.*, B, **34**, 187–220.

Cox, D.R. (1975). Partial likelihood. *Biometrika*, **62**, 269–76.

Cox, D.R. (1979). A note on the graphical analysis of survival data. *Biometrika*, **66**, 188–90.

Cox, D.R. (1980). Local ancillarity. *Biometrika,* **67**, 279–86.

Cox, D.R. (1983). A remark on censoring and surrogate response variables. *J. R. Statist. Soc.,* B 45, 391–3.

Cox, D.R. and Hinkley, D.V. (1968). A note on the efficiency of least-squares estimates. *J. R. Statist. Soc.,* B, **30**, 284–9.

Cox, D.R. and Hinkley, D.V. (1974). *Theoretical Statistics.* London: Chapman and Hall.

Cox, D.R. and Isham, V. (1980). *Point Processes.* London: Chapman and Hall.

Cox, D.R. and Lewis, P.A.W. (1966). *The Statistical Analysis of Series of Events.* London: Methuen.

Cox, D.R. and Snell, E.J. (1968). A general definition of residuals (with discussion). *J. R. Statist. Soc.,* B, **30**, 248–75.

Crowley, J. and Hu, M. (1977). Covariance analysis of heart transplant survival data. *J. Am. Statist. Assoc.,* **73**, 27–36.

Cuzick, J. (1982). Rank tests for association with right censored data. *Biometrika,* **69**, 351–64.

Daniels, H.E. (1944). The relation between measures of correlation in the universe of sample permutations. *Biometrika,* **32**, 129–35.

Darby, S.C. and Reissland, J.A. (1981). Low levels of ionizing radiation and cancer—are we understimating the risk? (with discussion). *J. R. Statist. Soc.,* A, **144**, 298–331.

David, H.A. and Moeschberger, M.L. (1978). *The Theory of Competing Risks.* London: Griffin.

Dempster, A.P., Laird, N.M. and Rubin, D.B. (1977). Maximum likelihood from incomplete data via the EM algorithm (with discussion). *J. R. Statist. Soc.,* B, **39**, 1–38.

DePriest, D.J. and Launer, R.L. (eds) (1983). *Reliability in the Acquisitions Process. Lecture Notes in Statistics,* vol. 4. New York: Marcel Dekker.

Dhillon, B.S. (1979). A hazard rate model. *IEEE Trans. Rel.,* **R-28**, 150.

Dhillon, B.S. (1981). Life distributions. *IEEE Trans. Rel.,* **R-30**, 457–60.

Doksum, K.A. (1974). Tailfree and neutral random probabilities and their posterior distributions. *Ann. Prob.,* **2**, 183–201.

Doksum, K.A. and Sievers, G.L. (1976). Plotting with confidence: graphical comparisons of two populations. *Biometrika,* **63**, 421–34.

Doksum, K.A. and Yandell, B.S. (1983). Properties of regression estimates based on censored survival data, in *A Festschrift for Erich L. Lehmann* (eds P.J. Bickel, K.A. Doksum and J.L. Hodges). New York: Wadsworth, pp. 140–56.

Efron, B. (1967). The two sample problem with censored data. *Proc. 5th Berkeley Symp.,* vol. 4, pp. 831–53.

Efron, B. (1977). The efficiency of Cox's likelihood function for censored data. *J. Am. Statist. Assoc.,* **72**, 557–65.

Efron, B. (1981). Censored data and the bootstrap. *J. Am. Statist. Assoc.*, **68**, 601–8.

Efron, B. and Hinkley, D.V. (1978). Assessing the accuracy of the maximum likelihood estimator: observed versus expected Fisher information (with discussion). *Biometrika*, **65**, 457–87.

Elandt-Johnson, R.C. and Johnson, N.L. (1980). *Survival Models and Data Analysis*. New York: Wiley.

Epstein, B. and Sobel, M. (1953). Life testing. *J. Am. Statist. Assoc.*, **48**, 486–502.

Farewell, V.T. (1979). An application of Cox's proportional hazard model to multiple infection data. *Appl. Statist.*, **28**, 136–43.

Farewell, V.T. and Cox, D.R. (1979). A note on multiple time scales in life testing. *Appl. Statist.*, **28**, 73–5.

Feigl, P. and Zelen, M. (1965). Estimation of exponential survival probabilities with concomitant information. *Biometrics*, **21**, 826–38.

Ferguson, T.S. (1973). A Bayesian analysis of some nonparametric problems. *Ann. Statist.*, **1**, 209–30.

Ferguson, T.S. and Phadia, E.G. (1979). Bayesian nonparametric estimation based on censored data. *Ann. Statist.*, **7**, 163–86.

Fisher, R.A. (1935). The mathematical distributions used in the common tests of significance. *Econometrica*, **3**, 353–65.

Fisher, R.A. and Tippett, L.H.C. (1928). Limiting forms of the frequency distribution of the largest or smallest member of a sample. *Proc. Camb. Phil. Soc.*, **24**, 180–90.

Fisher, R.A. and Yates, F. (1963). *Statistical Tables for Biological, Agricultural and Medical Research*, 6th edn. London: Longman.

Fleming, T.R., O'Fallon, J.R., O'Brien, P.C. and Harrington, D.P. (1980). Modified Kolmogorov–Smirnov test procedures with application to arbitrarily right censored data. *Biometrics*, **36**, 607–26.

Folks, J.L. and Chhikara, R.S. (1978). The inverse Gaussian distribution and its statistical application—a review (with discussion). *J. R. Statist. Soc.*, B, **40**, 263–89.

Fraser, D.A.S. (1968). *The Structure of Inference*. New York: Wiley.

Fraser, D.A.S. (1979). *Inference and Linear Models*. New York: McGraw-Hill.

Gail, M.H. (1972). Does cardiac transplantation prolong life? A reassessment. *Ann. Internal Med.*, **76**, 815–17.

Gail, M.H. (1975). A review and critique of some models used in competing risk analysis. *Biometrics*, **31**, 209–22.

Gail, M.H., Lubin, J.H. and Rubinstein, L.V. (1981). Likelihood calculations for matched case-control studies and survival studies with tied death times. *Biometrika*, **68**, 703–7.

Gehan, E.A. (1965). A generalized Wilcoxon text for comparing arbitrarily single-censored samples. *Biometrika*, **52**, 203–23.

Gilbert, J.P. (1962). Random censorship. Unpublished Ph.D. dissertation, University of Chicago.

Gill, R.D. (1980). *Censoring and Stochastic Integrals. Mathematical Centre Tracts, no.* 124. Amsterdam: Mathematische Centrum.

Gillespie, M.J. and Fisher, L. (1979). Confidence bands for the Kaplan–Meier survival curve estimate. *Ann. Statist.,* **7**, 920–4.

Glasser, M. (1965). Regression analysis with dependent variable censored. *Biometrics,* **21**, 300–7.

Glasser, M. (1967). Exponential survival with covariance. *J. Am. Statist. Assoc.,* **62**, 561–8.

Greenwood, M. (1926). The errors of sampling of the survivorship tables, in *Reports on Public Health and Statistical Subjects,* no. 33. London: HMSO. Appendix 1.

Gross, A.J. and Clark, V.A. (1975). *Survival Distributions: Reliability Applications in the Biomedical Sciences.* New York: Wiley.

Hall, W.J. and Wellner, J.A. (1980). Confidence bands for a survival curve from censored data. *Biometrika,* **67**, 133–43.

Harrington, D.P. and Fleming, T.R. (1982). A class of rank test procedures for censored survival data. *Biometrika,* **69**, 553–66.

Healy, M.J.R. and Westmacott, M. (1956). Missing values in experiments analysed on automatic computers. *Appl. Statist.,* **5**, 203–6.

Holt, J.D. and Prentice, R.L. (1974). Survival analysis in twin studies and matched pair experiments. *Biometrika,* **61**, 17–30.

Hougaard, P. (1984). Life table methods for heterogeneous populations: distributions describing the heterogeneity. *Biometrika,* **71**, to appear.

Howard, S.V. (1972). Contribution to discussion of paper by D.R. Cox. *J. R. Statist. Soc.,* B **34**, 210–11.

Irwin, J.O. (1949). The standard error of an estimate of expectation of life, with special reference to expectation of tumourless life in experiments with mice. *J. Hyg. Camb.,* **47**, 188–9.

Johansen, S. (1978). The product limit estimator as maximum likelihood estimator. *Scand. J. Statist.,* **5**, 195–9.

Johnson, N.L. and Kotz, S. (1970). *Distributions in Statistics. Continuous Univariate Distributions* (2 vols). Boston: Houghton Mifflin.

Jørgensen, B. (1982). *Statistical Properties of the Generalized Inverse Gaussian Distribution. Lecture Notes in Statistics,* vol. 9. New York: Springer Verlag.

Kalbfleisch, J.D. (1974). Some efficiency calculations for survival distributions. *Biometrika,* **61**, 31–8.

Kalbfleisch, J.D. (1978). Non-parametric Bayesian analysis of survival time data. *J. R. Statist. Soc.,* B **40**, 214–21.

Kalbfleisch, J.D. and Mackay, R.J. (1978). Remarks on a paper by Cornfield and Detre. *J. R. Statist. Soc.,* B **40**, 175–7.

Kalbfleisch, J.D. and McIntosh, A.A. (1977). Efficiency in survival distributions with time-dependent covariables. *Biometrika*, **64**, 47–50.

Kalbfleisch, J.D. and Prentice, R.L. (1973). Marginal likelihoods based on Cox's regression and life model. *Biometrika*, **60**, 267–78.

Kalbfleisch, J.D. and Prentice, R.L. (1980). *The Statistical Analysis of Failure Time Data*. New York: Wiley.

Kaplan, E.L. and Meier, P. (1958). Nonparametric estimation from incomplete observations. *J. Am. Statist. Assoc.*, **53**, 457–81.

Kay, R. (1977). Proportional hazard regression models and the analysis of censored survival data. *Appl. Statist.*, **26**, 227–37.

Kay, R. (1979). Some further asymptotic efficiency calculations for survival data regression models. *Biometrika*, **66**, 91–6.

Kendall, M.G. (1938). A new measure of rank correlation. *Biometrika*, **30**, 81–93.

Lagakos, S.W. (1979). General right censoring and its impact on the analysis of survival data. *Biometrics*, **35**, 139–56.

Lagakos, S.W. (1981). The graphical evaluation of explanatory variables in proportional hazard regression models. *Biometrika*, **68**, 93–8.

Lagakos, S.W. and Williams, J.S. (1978). Models for censored survival analysis: a cone class of variable-sum models. *Biometrika*, **65**, 181–9.

Langberg, N., Proschan, F. and Quinzi, A.J. (1981). Estimating dependent life lengths, with applications to the theory of competing risks. *Ann. Statist.*, **9**, 157–67.

Laslett, G.M. (1982). The survival curve under monotone density constraints with applications to two-dimensional line segment processes. *Biometrika*, **69**, 153–60.

Lawless, J.F. (1982). *Statistical Models and Methods for Lifetime Data*. New York: Wiley.

Liddell, F.D.K., McDonald, J.C. and Thomas, D.C. (1977). Methods of cohort analysis. Appraisal by application to asbestos mining (with discussion). *J. R. Statist. Soc.*, A **140**, 469–91.

Liu, P.Y. and Crowley, J. (1978). Large sample theory of the MLE based on Cox's regression model for survival data. University of Wisconsin, Technical Report.

Louis, T.A. (1982). Finding the observed information matrix when using the EM algorithm. *J. R. Statist. Soc.*, B **44**, 226–33.

Lustbader, E.D. (1980). Time-dependent covariates in survival analysis. *Biometrika*, **67**, 697–8.

McCullagh, P. (1980). Regression models for ordinal data (with discussion). *J. R. Statist. Soc.*, B **42**, 109–42.

Mann, N.R., Shafer, R.E. and Singpurwalla, N.D. (1974). *Methods for Statistical Analysis of Reliability and Life Data*. New York: Wiley.

Mantel, N. (1966). Evaluation of survival data and two new rank order statistics arising in its consideration. *Cancer Chemother. Rep.*, **50**, 163–70.

Mantel, N. and Byar, D.P. (1974). Evaluation of response time data involving transient states: an illustration using heart-transplant data. *J. Am. Statist. Assoc.*, **69**, 81–6.

Mantel, N. and Haenszel, W. (1959). Statistical aspects of the analysis of data from retrospective studies of disease. *J. Nat. Cancer Inst.*, **22**, 719–48.

Marshall, A.W. and Olkin, I. (1967). A multivariate exponential distribution. *J. Am. Statist. Assoc.*, **62**, 30–44.

Mehrotra, K.C., Michalek, J.E. and Mihalko, D. (1982). A relationship between two forms of linear rank procedures for censored data. *Biometrika*, **69**, 674–6.

Meier, P. (1975). Estimation of a distribution function from incomplete observations, in *Perspectives in Probability and Statistics* (ed. J. Gani). London: Academic Press, pp. 67–87.

Mendenhall, W. and Hader, R.J. (1958). Estimation of parameters of mixed exponentially distributed failure time distributions from censored life test data. *Biometrika*, **45**, 504–20.

Miké, V. and Stanley, K.E. (eds) (1982). *Statistics in Medical Research*. New York: Wiley.

Miller, R.G. (1976). Least squares regression with censored data. *Biometrika*, **63**, 449–64.

Miller, R.G. (1981). *Survival Analysis*. New York: Wiley.

Miller, R.G. and Halpern, J. (1982). Regression with censored data. *Biometrika*, **69**, 521–31.

Moeschberger, M.L. (1974). Life tests under dependent competing causes of failure. *Technometrics*, **16**, 39–47.

Moore, R.J. (1982). Derivatives of the incomplete gamma integral. Algorithm AS187. *Appl. Statist.*, **31**, 330–3.

Nádas, A. (1970). On estimating the distribution of a random vector when only the smallest coordinate is observable. *Technometrics*, **12**, 923–4.

Nádas, A. (1971). The distribution of the identified minimum of a normal pair determines the distribution of the pair. *Technometrics*, **13**, 201–2.

Nelson, W. (1969). Hazard plotting for incomplete failure data. *J. Qual. Technol.*, **1**, 27–52.

Nelson, W. (1972). Theory and applications of hazard plotting for censored failure data. *Technometrics*, **14**, 945–65.

Nelson, W. (1982). *Applied Life Data Analysis*. New York: Wiley.

Oakes, D. (1972). Contribution to discussion of paper by D.R. Cox. *J. R. Statist. Soc.*, B **34**, 208.

Oakes, D. (1977). The asymptotic information in censored survival data. *Biometrika*, **64**, 441–8.

Oakes, D. (1981). Survival times: aspects of partial likelihood (with discussion). *Int. Statist. Rev.*, **49**, 235–64.

Oakes, D. (1982a). A concordance test for independence in the presence of censoring. *Biometrics*, **38**, 451–5.

Oakes, D. (1982b). A model for association in bivariate survival data. *J. R. Statist. Soc.*, B **44**, 414–22.

Orchard, T. and Woodbury, M.A. (1972). A missing information principle: theory and applications. *Proc. 6th Berkeley Symp.*, vol. 1, pp. 697–715.

Pearson, E.S. and Hartley, H.O. (1966). *Biometrika Tables for Statisticians*, 3rd edn. London and Cambridge: Cambridge University Press.

Pereira, B. de. B. (1978). Tests and efficiencies of separate regression models. *Biometrika*, **65**, 319–27.

Peterson, A.V. (1977). Expressing the Kaplan–Meier estimate as a function of empirical subsurvival functions. *J. Am. Statist. Assoc.*, **72**, 854–8.

Peto, R. (1972). Contribution to discussion of paper by D.R. Cox. *J. R. Statist. Soc.*, B **34**, 205–7.

Peto, R. and Lee, P. (1973). Weibull distributions for continuous carcinogenesis experiments. *Biometrics*, **29**, 457–70.

Peto, R. and Peto, J. (1972). Asymptotically efficient rank invariant test procedures (with discussion). *J. R. Statist. Soc.*, A **135**, 185–206.

Peto, R., Pike, M.C., Armitage, P., Breslow, N.E., Cox, D.R., Howard, S.V., Mantel, N., McPherson, K., Peto, J. and Smith, P.G. (1976). Design and analysis of randomized clinical trials requiring prolonged observation of each patient. I. Introduction and design. *Br. J. Cancer*, **34**, 585–612.

Peto, R., Pike, M.C., Armitage, P., Breslow, N.E., Cox, D.R., Howard, S.V., Mantel, N., McPherson, K., Peto, J. and Smith, P.G. (1977). Design and analysis of randomized clinical trials requiring prolonged observation of each patient. II. Analysis and examples. *Br. J. Cancer*, **35**, 1–39.

Peto, R., Pike, M.C., Day, N.E., Gray, R.G., Lee, P.N., Parish, S., Peto, J., Richards, S. and Wahrendorf, J. (1980). Guidelines for simple, sensitive significance tests for carcinogenic effects in long-term animal experiments, in *Long-Term and Short-Term Screening Assays for Carcinogens: A Critical Appraisal*. Lyon: IARC.

Pierce, D.A., Stewart, W.H. and Kopecky, K.J. (1979). Distribution free regression analysis of grouped survival data. *Biometrics*, **35**, 785–93.

Pike, M.C. (1966). A suggested method of analysis of a certain class of experiments in carcinogenesis. *Biometrics*, **22**, 142–61.

Plackett, R.L. (1965). A class of bivariate distributions. *J. Am. Statist. Assoc.*, **60**, 516–22.

Prentice, R.L. (1978). Linear rank tests with right censored data. *Biometrika*, **65**, 167–79.

Prentice, R.L. and Gloeckler, L.A. (1978). Regression analysis of grouped survival data with application to breast cancer data. *Biometrics*, **34**, 57–67.

Prentice, R.L. and Kalbfleisch, J.D. (1979). Hazard rate models with covariates. *Biometrics*, **35**, 25–39.

Prentice, R.L. and Marek, P. (1979). A qualitative discrepancy between censored data rank tests. *Biometrics*, **35**, 861–7.

Prorok, P.C. (1976). The theory of periodic screening I: lead time and proportion detected. *Adv. Appl. Prob.*, **8**, 127–43.

Rao, C.R. (1973). *Linear Statistical Inference and its Applications*, 2nd edn. New York: Wiley.

Reid, N. (1981a). Influence functions for censored data. *Ann. Statist.*, **9**, 78–92.

Reid, N. (1981b). Estimating the median survival time. *Biometrika*, **68**, 601–8.

Schmee, J. and Hahn, G.J. (1979). A simple method for regression analysis with censored data. *Technometrics*, **21**, 417–32.

Schneider, B.E. (1978). Trigamma function. Algorithm AS121. *Appl. Statist.*, **27**, 97–9.

Seal, H.L. (1977). Studies in the history of probability and statistics, XXXV. Multiple decrements or competing risks. *Biometrika*, **64**, 429–39.

Self, S.G. and Prentice, R.L. (1982). Commentary on Andersen and Gill's 'Cox's regression model for counting processes: a large sample study'. *Ann. Statist.*, **10**, 1121–4.

Sen, P.K. (1981). The Cox regression model, invariance principles for some induced quantile processes and some repeated significance tests. *Ann. Statist.*, **9**, 109–21.

Shahani, A.K. and Crease, D.M. (1977). Towards models of screening for early detection of disease. *Adv. Appl. Prob.*, **9**, 665–80.

Shapiro, S. (1977). Evidence on screening for breast cancer from a tandomized trial. *Cancer*, **39**, 2772–82.

Sukhatme, P.V. (1937). Tests of significance for samples of the χ^2 population with two degrees of freedom. *Ann. Eugen.*, **8**, 52–6.

Sundberg, R. (1974). Maximum likelihood theory for incomplete data from an exponential family. *Scand. J. Statist.*, **1**, 49–58.

Sundberg, R. (1976). An iterative method for solution of the likelihood equations for incomplete data from exponential families. *Comm. Statist.*, B **5**, 55–64.

Susarla, V. and Van Ryzin, J. (1976). Nonparametric Bayesian estimation of survival curves from incomplete observations. *J. Am. Statist. Assoc.*, **71**, 897–902.

Thomas, D.C. (1981). General relative risk models for survival time and matched case-control analysis. *Biometrics*, **37**, 673–86.

Tsiatis, A.A. (1975). A nonidentifiability aspect of the problem of competing risks. *Proc. Nat. Acad. Sci. USA*, **72**, 20–2.

Tsiatis, A.A. (1981). A large sample study of Cox's regression model. *Ann. Statist.*, **9**, 93–108.

Turnbull, B.W. (1974). Nonparametric estimation of a survivorship function with doubly censored data. *J. Am. Statist. Assoc.*, **69**, 169–73.

Turnbull, B.W. (1976). The empirical distribution function with arbitrarily grouped, censored and truncated data. *J. R. Statist. Soc.*, B **38**, 290–5.

Turnbull, B.W., Brown, B.W. and Hu, M. (1974). Survivorship analysis of heart transplant data. *J. Am. Statist. Assoc.*, **69**, 74–80.

Vaeth, M. (1979). A note on the behaviour of occurrence-exposure rates when the survival distribution is not exponential. *Scand. J. Statist.*, **6**, 77–80.

Vaupel, J.W. Mantom, K.G. and Stallard, E. (1979). The impact of heterogeneity in individual frailty on the dynamics of mortality. *Demography*, **16**, 439–54.

Vardi, Y. (1982a). Nonparametric estimation in the presence of length bias. *Ann. Statist.*, **10**, 616–20.

Vardi, Y. (1982b). Nonparametric estimation in renewal processes. *Ann. Statist.*, **10**, 772–85.

Weibull, W. (1939a). *A Statistical Theory of the Strength of Materials. Ingeniörs Vetenskaps Akademien Handlingar*, no. 151.

Weibull, W. (1939b). *The Phenomenon of Rupture in Solids, Ingeniörs Vetenskaps Akademien Handlingar*, no. 153.

Weier, D.R. and Basu, A.P. (1980). An investigation of Kendall's tau modified for censored data with applications. *J. Statist. Plan. Infer.*, **4**, 381–90.

Widder, D.V. (1946). *The Laplace Transform*. Princeton, NJ: Princeton University Press.

Wild, C.J. (1983). Failure time models with matched data. *Biometrika*, **70**, 633–41.

Wilk, M.B. and Gnanadesikan, R. (1968). Probability plotting methods for the analysis of data. *Biometrika*, **55**, 1–17.

Williams, J.S. and Lagakos, S.W. (1977). Models for censored survival analysis: constant-sum and variable-sum models. *Biometrika*, **64**, 215–24.

Wolynetz, M.S. (1979). Maximum likelihood estimation in a linear model from confined and censored normal data. Algorithm AS139. *Appl. Statist.*, **28**, 195–206.

Wu, C.F.J. (1983). On the convergence properties of the EM algorithm. *Ann. Statist.*, **11**, 95–103.

Yates, F. (1933). The analysis of replicated experiments when the field results are incomplete. *Empire J. Exptl Agric.*, **1**, 129–42.

Zelen, M. and Dannemiller, M.C. (1961). The robustness of life testing procedures derived from the exponential distribution. *Technometrics*, **3**, 29–49.

Zelen, M. and Feinleib, N. (1969). On the theory of screening for chronic diseases. *Biometrika*, **56**, 601–14.

Author index

195

Subject index